EXPORT OR DIE

*'Britain exports or dies.'*

David Mellor MP,
former Minister of State for Foreign and Commonwealth Affairs.
(*Inquiry into Exports of Defence Equipment and
Dual-Use Goods to Iraq*,
Hearings in the presence of
The Right Honourable Lord Justice Scott, Day 25, p. 58.)

# Export or Die

## Britain's Defence Trade with
## Iran and Iraq

## Davina Miller

CASSELL

**Cassell**
Wellington House, 125 Strand, London WC2R 0BB
127 West 24th Street, New York, NY 10011

First published 1996

**British Library Cataloguing-in-Publication Data**
A catalogue record for this book is available from the British Library.

ISBN 0-304-33852-4 (hardback)
    0-304-33853-2 (paperback)

Designed and typeset by Kenneth Burnley at Irby, Wirral, Cheshire.
Printed and bound in Great Britain by Biddles Limited, Guildford and
King's Lynn.

# Contents

# Acknowledgements

This book began as a doctoral thesis. The impetus to study Britain's arms sales began with my concerns about the consequences of the defence trade, but the original idea came from Sharam Taromsari, who was also an excellent teacher of Middle Eastern politics. Colleagues at the Department of Peace Studies at Bradford were supportive throughout the completion of this work, but, in particular, Paul Rogers and Malcom Chalmers gave excellent advice. Lawrence Freedman steered me towards what I was in danger of overlooking, and his, and Robert McKinlay's, comments on the thesis helped shape the book. Discussions with Mark Phythian were invaluable in organizing my thoughts. Above all, Martin Edmonds, my PhD supervisor, was an unflinching critic and a tireless editor. I owe a large debt to the Rt Hon. Alan Clark, the Rt Hon. Sir Adam Butler and Sir Hal Miller who spoke to me at length and with candour. Gerald James, Chris Cowley, Paul Henderson and particularly Kevin Robinson gave me generous and indispensable assistance. I am also grateful to those who spoke to me anonymously. The responsibility for any errors, however, remains mine alone.

Jane Greenwood, Steve Cook and Kenneth Burnley at Cassell have been enormously helpful. My sister Denise came up with the title and took on the majority of the word processing, which she did with absolute dedication. I am also grateful to her and to my brother Andrew for reading the various chapters thrust upon them. I am indebted to all my family and friends for their understanding but especially to Sheila Kerr.

DAVINA MILLER

# Foreword

I recall with crystal clarity the day, in 1980, when I heard a radio report that hostilities had broken out between Iran and Iraq. Having absorbed the fact, I thought little more about it despite seeing, in the succeeding eight years, the reports of missile strikes, gas attacks and massive casualties: it was a long way away and had nothing to do with my country or me.

In April 1990, after the recovery of tubes at Teeside docks by Customs and Excise, I was asked to act for Christopher Cowley, the supergun's project manager. Both from what he told me, and from the response of the prosecuting authorities to what he was saying, it was clear that my long-held belief that the Iran–Iraq War was nothing to do with Britain, or with me, was a great self-deception. After Christopher Cowley, I represented Paul Henderson of Matrix Churchill, Gerald James (Astra's Chairman), and Paul Grecian of Ordtech. All four had different stories to tell me, but the common thread was the fact that the Government operated a policy that was at odds with what it said was the case.

It is now a matter of history that Paul Henderson's acquittal in November 1992 was the catalyst for the Scott Inquiry, set up in the political maelstrom of the days following the abandonment of the prosecution. The Inquiry was given carefully and narrowly drawn terms of reference. Rather than conducting an investigation into Britain's arms trade with Iraq, Sir Richard Scott concentrated on the minutiae of Government procedures based on the documents released to him.

The real problem with the Scott Inquiry was that it was always going to be a political event rather than anything else: it was destined to be hijacked for political purposes, with politicians and commentators finding sections to prove their disparate points of view. The effect was that more heat than light was generated. But, in all of this, we should not let slip the opportunity of learning from both the events which led to the Inquiry and from the information it subsequently uncovered. Politicians and journalists might have moved on, but the issues and implications remain.

The value of Davina Miller's book is that it places the British Government's activities in relation to Iran and Iraq in their true context, unrestrained by narrowly drawn terms of reference. We are fortunate to have in this book a scholarly, dispassionate and yet eminently readable analysis of Britain's policy towards Iran and Iraq. The very great effort and skill which went into the writing will be clear to any reader who, in turn, will be rewarded by a lucid and insightful inquiry into fundamental issues of Britain's international relations and their impact on domestic politics.

KEVIN ROBINSON
*Sheffield, April 1996*

# Introduction

What defence equipment Britain did or did not supply to Iran and Iraq during the 1980s has continued to be a matter of contention. Similarly, the political and constitutional issues raised, by media coverage, accounts by participants, public inquiry and Government denial, have yet to be resolved. However, both debates have been conducted without a proper understanding of Britain's policy towards Iran and Iraq and the motives which lay behind it. This book is an attempt to provide that analysis.

The controversy began with the outbreak of war between Iran and Iraq in 1980. The British Government, having declared its neutrality, said it would not sell lethal equipment to either participant. In 1984 this policy was apparently strengthened by the introduction of the following Guidelines:

1. We should maintain our consistent refusal to supply lethal equipment to either side.
2. Subject to that overriding consideration, we should attempt to fulfil existing contracts and obligations.
3. We should not, in future, approve orders for any defence equipment which, in our view, would significantly enhance the capability of either side to prolong or exacerbate the conflict.
4. In line with this policy, we should continue to scrutinise rigorously all applications for export licences for the supply of defence equipment to Iran and Iraq.[1]

The British Government instituted official machinery to execute the policy and constantly asserted in Parliament that, in contrast to other countries, it did not sell arms to either party. In 1988, following the ceasefire between Iran and Iraq, the third Guideline was amended to read:

We should not in future approve orders for any defence equipment which, in our view, would be of direct and significant assistance to either country in the conduct of offensive operations in breach of the ceasefire.[2]

Following Khomeini's death sentence upon Salman Rushdie in February 1989, defence sales for Iran were judged against the original, more restrictive, Guidelines.

Iraq's procurement of arms-making equipment and components for non-conventional weapons drew media attention in the late 1980s, while allegations that British arms had been clandestinely sold to both Iraq and Iran periodically surfaced throughout the decade. Thus, for the media and the public, there were two issues. The first was the true nature of the UK's arms export policy towards two states with very poor human rights records and engaged in an extremely bloody conflict. The second was the UK's policy towards Iraq's procurement network.

After reports, in 1990, that two British companies had exported parts of a supergun, having taken advice from officials at the Department of Trade and Industry that export licences were not required, the House of Commons examined Project Babylon and other long-range gun projects in which British companies were involved.[3] The Trade and Industry Committee concluded that, in essence, the affair could be explained by reference to miscommunication within and between government departments. The Committee was unable to examine the role of the Secret Intelligence Service (SIS) – specifically allegations that the SIS had been informed at an early stage of the project – because papers and witnesses were not forthcoming.

Continuing press reports of the UK's defence trade with Iraq in the 1980s were imbued with political sensitivity after Iraq's invasion of Kuwait in August 1990 as British forces faced an enemy equipped with British military technology and as questions were asked about the UK's responsibility in nurturing Iraqi aggression with trade and with credit.

On 16 October 1990, Paul Henderson and two fellow-directors of Matrix Churchill were arrested for breaches of export law. At their trial two years later, the judge overruled Public Interest Immunity (PII) Certificates, signed by four Ministers, and released to the defence documents which showed that, in spite of the companies declaring the end use to be 'general engineering', the Government knew that the machine tools were to be used to make Iraqi weapons. The Government knew this because Paul Henderson had told the SIS that that was the case. Alan Clark conceded in the witness box that company and Government had colluded over the declared end use of the machine tools. The prosecution dropped the case and the Scott Inquiry was set up by the Government.

The Inquiry into Exports of Defence Equipment and Dual-Use Goods to Iraq lasted three years; received evidence from more than 200 people; gathered some 200,000 pages of documents; and allowed the public to witness the examination of Ministers and officials. The resulting Report, at 1,806 pages, raises as many issues as it resolves.[4]

Against all the official foreboding of Sir Richard's inability to understand the real world of diplomacy, trade and other arms suppliers, his Report stated that, 'Government's *actual* policy was well capable of being supported in argument' (emphasis added).[5] His criticisms revolved around the questions of Ministers' accountability to Parliament and failures in administrative systems or errors by civil servants.

The Report noted that in 'example after example' Ministers had failed to honour Questions of Procedure in not discharging:

> the duty to give Parliament . . . as full information as possible about the policies, decisions and actions of the Government, and not to deceive or mislead Parliament and the public.[6]

In particular, the Minister of State at the Foreign and Commonwealth Office (FCO), William Waldegrave, continued, with the agreement of his counterparts in the Ministry of Defence (MOD) and the Department of Trade and Industry (DTI), to assert the original Guidelines as Government policy after their amendment in late 1988:

> Parliament and public were designedly led to believe that a stricter policy towards non-lethal defence exports and dual-use exports to Iraq was being applied than was in fact the case.[7]

The Attorney General, Sir Nicholas Lyell, was also singled out as being 'personally at fault' in the use of PII Certificates in the Matrix Churchill case.[8]

Civil servants were blamed for their actions in failing, on a number of occasions, to share information with Ministers and with each other. In finding overwhelming evidence of official knowledge of Jordan's diversion of British defence goods to Iraq, Sir Richard asserted that Government policy was 'undermined'. If, as Government witnesses argued, that was the 'price' of maintaining a very necessary relationship with Jordan, then that decision should 'have been specifically put to and taken by Ministers'. Sir Richard found no evidence that that was ever done.[9]

The Report found officials at fault for not reflecting, in their submissions to Ministers, specific intelligence about the end use of one batch of machine tools in the Iraqi missile programme. By implication, Sir Richard did not accept their collective explanation that intelligence reports were forgotten about. As a result, Ministers took the decision to allow the export 'on a false footing'.[10] Alan Barrett, the Deputy to the Head of the Defence Export Services Secretariat (DESS), was blamed in the Report for not telling the Minister for Defence Procurement (MinDP) about both intelligence regarding an earlier batch of machine tools and a dispute between the SIS and the Defence Intelligence Service (DIS) over the use of that intelligence.[11] Officials were also criticized for failing to share information about the supergun.[12]

Sir Richard went on to accept the Cabinet Secretary's distinction between Ministerial accountability and Ministerial responsibility. While the Minister remains the person who must give an account for the work of his/her Department to Parliament:

> A Minister should not be held to blame or required to accept personal criticism unless he has some personal responsibility for or some personal involvement in what has occurred.[13]

The renewed discussion of the implications of the affair for both the Conservative Government and the British Constitution has taken place without any understanding – and, indeed, in many cases with a good deal of misunderstanding – of both the policy and the forces behind that policy. Without this, it is impossible to draw lessons from the affair for the British system of government; for accountability, open government, and the probity of the civil service. For example, the responsibility for policy between Ministers and civil servants cannot be established unless both the true purpose and operation of the Guidelines are properly understood. Whether the diversion of British arms by other countries can be understood either as failures in control mechanisms or as complementary to the Government's aims depends on the real nature of policy. Moreover, systems of government exist within wider economic and political contexts: the actions of the Executive are not comprehensible without an understanding of the environment in which decisions were taken since defence exports are not merely an end in themselves, but are an expression of other policies.

Constitutional reform might not be the solution to the democratic concerns raised by the Scott Report so much as a thoroughgoing debate about the nature of Britain's international relations. The decision-makers' perception that arms sales are central to the pursuit of influence with developing states competes, with constitutional or moral weakness, as an explanation for the failure of Ministers to give information to Parliament. Thus, it might be most appropriate to re-evaluate this perception, especially in the light of evidence of the reverse influence which arms exports bring. The Home Office attempt, in 1996, to expel the dissident and agitator, Mohammed al-Mas'ari, from Britain in response to pressure from his native Saudi Arabia is a case in point.

Defence exports have also to be seen in their economic context. In 1985, the House of Lords Select Committee on Overseas Trade summed up Britain's economic predicament:

> the continuing deficit in the balance of trade in manufactures is a symptom of the decline in Britain's manufacturing capacity on the one hand and of the poor competitiveness of important areas of manufacturing on the other. Unless the climate is changed so that the manufacturing base is enlarged and steps are taken to ensure that import penetration is combatted and that manufactured exports are stimulated, as the oil revenues diminish the country will experience adverse effects which will worsen with time.[14]

Arms sales represented a success in a decade which saw exports rise at less than half the rate of imports. More significantly, the major economic motive to sell defence equipment to Iran and Iraq was to maintain ordinary commercial relationships with the two countries and with Iraq's Arab supporters. Officials were reminded that, when applying the Guidelines, they should not 'prejudice civil trade'.[15] As a former Head of the Middle East Department commented:

> if Her Majesty's Government refused to supply a particular item of equipment to one or other belligerent, would that government retaliate by cancelling contracts with a range of entirely unrelated British suppliers? At the time this was seen as a serious threat.[16]

The Matrix Churchill documents and those read into the record of the Scott Inquiry make clear a perception held by officials and Ministers that a refusal to sell defence equipment could have implications for civil trade. To illustrate, in spite of American and Arab pressure to cease supplying Iran, especially with military spares, the UK continued to do so for fear of the consequences for Anglo–Iranian commercial relations.[17] In these respects, it might be wrong to dismiss the affair as simply attributable to a blind pursuit of commercialism and a corresponding breakdown in ethical standards. To understand Britain's policy towards defence exports to Iran and Iraq, one needs to distinguish the sorts of pressures, internal and external, which all governments confront. It should be remembered that the Labour Government of the 1970s sold the repressive Shah of Iran large quantities of advanced weaponry.

Paul Henderson, the Managing Director of Matrix Churchill, felt he had no apology to make for trading with Iraq: selling machine tools that will make munitions components is, he said, 'an inescapable fact of industrialisation'.[18] Arms and arms-making technology are *controlled* because they represent the diffusion of strength in a competitive system. But, arms are *transferred* because states inhabit the same competitive environment. International rivalry, to state the obvious, has economic as well as political and military dimensions. This is the central problem that confronts the decision-maker: it is simply the 'guns or butter' predicament in another guise. In the absence of any international control regime, if the UK does not supply to a particular country, there are other states who will. Revenue is lost without the achievement of any objective. In other words, the problem can be seen as structural as much as moral.

This is not to say that there are not choices to be made, that the trade in arms is an imperative imposed by the nature of the international system. But, if exports of weapons and dual-use technology are to be restrained, one of two things would have to happen first. Either an international consensus needs to be developed and enforced about particular countries or particular types of arms; or the British public has to articulate its concerns about avoiding the sale of weapons to aggressive regimes. Arms control regimes thus far have reflected the interests of the powerful rather than the moral sensibilities of the international community – control regimes exist only in respect of non-conventional arms and missiles – and governments come and go on the basis of public perceptions of economic performance.[19] The Matrix Churchill and Ordnance Technologies directors were prosecuted by a Government unable – with the possible exception of Alan Clark – to explain its perceived dilemmas to its citizens. Perhaps this says as much as anything else in the affair about the health of British democracy.

As for the future, the Scott Report made four recommendations that affect defence sales policy. First, export licensing legislation should be overhauled in two respects: Parliament ought to exert more influence over the imposition of controls; and, preceded by an open debate, the purposes of export control should be made narrow and precise. Second, in order to discharge their constitutional duties under the principle of Ministerial responsibility, the Government should give more information to Parliament about arms sales. Third, consideration should be given to removing defence export control from the DTI to the MOD. Fourth, steps should be taken to ensure the proper dissemination of intelligence relevant to export licensing decisions.[20]

The benefits of selling arms are far from clear. There is a lack of hard data upon which to judge Government claims about the positive effects for employment, the balance of trade, the offsetting of domestic procurement costs and the maintenance of capacity in the defence manufacturing base. On the other hand, some costs are clear. The provision of credit was judged by Defence Attachés to be the most important factor in securing arms sales:[21] the financial burden in terms of unpaid debts to the Export Credits Guarantee Department (ECGD) is thus one such cost. Another is the estimated £81 million spent each year in promoting weapons exports.[22] The international political costs of arms sales include arming potential enemies and alienating future governments currently disguised as dissidents and troublemakers. But there are also domestic political costs, as evidenced in the case of Mohammed al-Mas'ari. British sovereignty was undermined by the reverse influence which arms sales bring.

Openness about arms exports might help to properly define their costs and benefits, and the debate about British export controls clarify perceptions of what our relationships with the world are and ought to be. To what extent we are likely to have greater freedom of information in this respect, given the preoccupations of the past and the political culture of the present, remains to be seen. Sir Richard's desire that considerations of foreign policy be removed from decision-making about defence and dual-use exports is highly unlikely to be satisfied. As the analysis of Britain's arms sales policy towards Iran and Iraq makes clear, these exports are a means of conducting British foreign policy.

However, inflexible export controls would have the advantage of removing the recipient's ability to influence the direction of Government policy, viz the successful Iraqi pressure to relax British restrictions in 1990 against the backdrop of Iraq's known procurement of weapons of mass destruction.

This book is not intended to be a commentary upon the Scott Report. Sir Richard has produced an account, as opposed to an explanation, of Britain's arms export policy towards Iraq. I have sought to give an analysis which includes both Iraq and Iran. Furthermore, while Sir Richard, in framing the narrative, drew conclusions solely from the documentation peculiar to any one particular stage in the story, this book examines both the direct and the circumstantial evidence. I have reached different judgements from Sir Richard regarding: senior Ministers' knowledge of the changing of the Guidelines on defence sales to Iran and Iraq; and Ministers' awareness of intelligence in the cases of the end use of machine tools, the involvement of British companies in manufacturing the supergun, and diversion. I have highlighted those different conclusions as they occur in the course of the book.

In examining all the evidence available to me, I do not emphasise muddle, or error by civil servants. Rather, the evidence suggests that the undiluted notion of Ministerial responsibility remains valid. Given these differences, I have quoted extensively. This might not be to everyone's taste, but it does allow the reader to see as much of the evidence as possible. I am able to do so because of the way Sir Richard conducted the hearings and presented the Report. No matter what the critics might have made of his conclusions, or lack of them, those who care to study the Report can judge the evidence for themselves.

I make no judgements about the morality, or the wisdom, of selling defence goods to Iran and Iraq in the context of both the war and their abhorrent domestic and international behaviour: these are judgements that I am no more qualified to make than the next person. The following chapter outlines the main conclusions of this study and sets them against previous assumptions about the arms trade. Chapter 2's main purpose is to establish the nature of the decision-making process. This is vital if responsibility for policy as between Ministers and civil servants is to be properly attributed, and is something the Scott Report did not do. Chapters 3 and 4 analyze overt defence exports policy, the Guidelines, and Chapter 5 examines covert sales to Iran and Iraq. Chapters 6, 7, 8 and 9 discuss Britain's policy towards the Iraqi procurement network. The book concludes with a discussion of the extent to which extrapolations can be made from this case study about the behaviour of other arms supplying countries.

# Guide to abbreviations

| | |
|---|---|
| ACDA | Arms Control and Disarmament Agency (US) |
| ACDD | Arms Control and Disarmament Department, FCO |
| ARV | Armoured Recovery Vehicle |
| AUSC | Assistant Under-Secretary Commitments |
| AWP | Arms Working Party |
| | |
| BAe | British Aerospace |
| BMARC | British Manufacture and Research Company |
| BNL | Banca Nazionale de Lavoro |
| Box 850 | Secret Intelligence Service |
| Box 500 | Security Service |
| BW | Biological Weapons |
| | |
| CIA | Central Intelligence Agency |
| COCOM | Co-ordinating Committee for Multilateral Strategic Controls |
| COMED | Commercial Management and Exports Department, FCO |
| CW | Chemical Weapons |
| CWC | Chemical Weapons Convention |
| | |
| DESO | Defence Exports Services Organisation |
| DESS | Defence Exports Services Secretariat |
| DESS 2 | Deputy to Head of DESS and Chair of MODWG |
| DIS | Defence Intelligence Staff |
| DTI | Department of Trade and Industry |
| | |
| ECGD | Export Credits Guarantee Department |
| ECO | Export Control Organisation |
| EGC | Export Guarantees Committee |
| EG(C)O | Export of Goods (Control) Order |

| | |
|---|---|
| ELA | Export Licence Application |
| ELB | Export Licensing Branch |
| ELU | Export Licensing Unit |
| | |
| FCO | Foreign and Commonwealth Office |
| | |
| GCC | Gulf Co-operation Council |
| GCHQ | Government Communications Headquarters |
| | |
| HDES | Head of Defence Export Services |
| | |
| IAEA | International Atomic Energy Agency |
| IDC | Interdepartmental Committee on Defence Sales to Iran and Iraq |
| IMS | International Military Services Limited |
| IMPOS | Iranian Military Procurement Offices |
| | |
| JIC | Joint Intelligence Committee |
| | |
| MBT | Main Battle Tank |
| MED | Middle East Department, FCO |
| MFT | Minister for Trade |
| MI5 | The Security Service |
| MinAF | Minister for the Armed Forces |
| MinDP | Minister for Defence Procurement |
| MOD | Ministry of Defence |
| MODWG | MOD Working Group, Iran/Iraq |
| MOU | Memorandum of Understanding |
| MTCR | Missile Technology Control Regime |
| MTTA | Machine Tools (Trade) Technologies Association |
| | |
| NAO | National Audit Office |
| NBC | Nuclear, Biological, Chemical |
| NEC | Nobel Explosives Company, ICI |
| NENAD | Near East and North Africa Department, FCO |
| NPDD | Non-Proliferation and Defence Department, FCO |
| NPT | Nuclear Non-Proliferation Treaty |
| NSG | Nuclear Suppliers Group |
| | |
| OD | Overseas and Defence Committee of the Cabinet |
| OGEL | Open General Export Licence |

| | |
|---|---|
| OIEL | Open Individual Export Licence |
| OT2/3 | Overseas Trade Division, DTI, responsible for export control until 1992 |
| | |
| PQ | Parliamentary Question |
| PRB | Pouderies Réunies de Belgique |
| PUS | Permanent Under-Secretary |
| PUSD | Permanent Under-Secretary's Department, FCO |
| | |
| RARDE | Royal Armaments Research and Development Establishment |
| REU | Restricted Enforcement Unit |
| RMD | Regional Marketing Desk, DESO |
| RMIPC | Release of Military Information Policy Committee |
| ROF | Royal Ordnance Factories |
| | |
| SECS | Security Exports Control Section, DTI |
| SEND | Science, Energy, Nuclear Department, FCO |
| SIPRI | Stockholm International Peace Research Institute |
| SIS | Secret Intelligence Service |
| SRC | Space Research Corporation |
| STU | Sensitive Technologies Unit |
| SXWP | Security Export Controls Working Party |
| | |
| TDG | Technology Development Group (Iraqi owned procurement organization) |
| TEG | Technology Engineering Group (Iraqi owned procurement organization) |
| TMG | TMG Engineering (Iraqi owned parent company of Matrix Churchill) |
| TRED | Trade Relations and Export Department, FCO |
| | |
| WGIP | Working Group on Iraqi Procurement |
| WMD | Weapons of Mass Destruction |

# Tables and diagrams

# 1

# The motives of
# arms exporting nations

Since the 1970s, there has been 'an explosion of studies'[1] about the defence trade but, because of the secrecy surrounding the sale of arms, understanding of its dynamics remains poor. Valuable sources for scholars have been established: the Stockholm International Peace Research Institute's (SIPRI) *Yearbook of World Armaments and Disarmament*; the US Arms Control and Disarmament Agency's (ACDA) *World Military Expenditures and Arms Transfers*; and the International Institute for Strategic Studies' *The Military Balance*.[2] But these, like much of the literature about the weapons trade, provide only raw statistics together with current trends. They should not be confused with, as Ralph Huitt has it, 'the basic descriptions of a phenomenon (that) are central to an explanation of it'.[3]

That these 'basic descriptions' are lacking is demonstrated by the urging of the need for empiricism by a prominent scholar in 1994:

> We should first establish what is going on: who does what in the pursuit of which interests and with what kind of consequences. This is the basis required for a successful search for underlying patterns.[4]

The general scarcity of data has had two consequences for the study of the arms trade: a focus upon the collection of information, a 'datafixation';[5] and a persistent uncertainty about the quality of analysis based upon partial and controvertible information. As to the causes of arms exporting, Klare argued:

> In most major deals . . . a combination of motives is involved and hence it is extremely difficult (without inside information) to isolate the decisive factors.[6]

Through sheer serendipity – from a succession of Government error, legal procedure and public inquiry – students of the arms trade now have a window upon decision-makers' motives. This book, in utilizing Government documents, as well as interviews with participants, is partially able to overcome the problem of data about the motives of suppliers. It is only a partial overcoming of the problem because: there is access only to a limited set of papers; with a few exceptions, these papers are carefully crafted or sanitized; and, moreover, the papers do not record the informal, the frank, discussions which precede the decision. Nonetheless, such access to the decision-making process is unprecedented. Given the uncertainty of scholarship about the aims of arms suppliers, this book investigates the evidence of the nature of economic and political forces. What is especially under-researched is the relationship between defence and civil trade, mentioned but unexplored in the existent literature because of the lack of data. This book attempts to examine that relationship in close detail.

This case study is particularly useful in two respects. First, Britain is a major supplier of arms. Between 1980 and 1990, SIPRI estimated that Britain was the fourth largest exporter of arms, behind the US, Soviet Union and France.[7] According to Government figures, arms exports rose from £1,074 million in 1980 to £1,980 million by 1990.[8] By 1986, Britain had acquired 20 per cent of the market, placing it second behind the US, according to the Government. This figure is largely accounted for by the sale of Tornado aircraft and other equipment to Saudi Arabia.[9] In common with France and Germany, Britain, in the 1980s, was dependent upon the Third World for more than 80 per cent of its arms exports.[10] Second, Iran and Iraq represent the most significant market for arms: in the 1980s, the Middle East accounted for more than 50 per cent of British and other countries' weapons sales.[11]

Because of these two factors, this case study can tell us something about the current arms trade, given that all suppliers, in the post-Cold War era, are said to be driven by the motives previously ascribed to countries such as Britain. The case of Britain's defence exports to Iran and Iraq also provides an opportunity to examine the transfer of illicit arms and dual-use goods. In the case of the latter, it permits an evaluation of the effectiveness of current international control regimes. Although primarily concerned with the supplier, this work also reveals something of the motives and behaviour of recipients.

It should, however, be noted that, as a case study, it has its limitations. Those limitations are not diminished by its being a somewhat narrow case study: it examines one supplier and one recipient region in a period of ten years. These restrictions are not chosen but imposed by the evidence available. Nonetheless, the strengths of this work lie in the quantity and quality of its evidence. In brief, the main contentions here are: that it is not sufficient to attempt to explain the arms sales of non-superpower suppliers simply in terms of the political desire to maintain the defence manufacturing base; and it is simplistic to see the arms trade as a purely commercial affair. The arms trade needs to be located both within the broader economic system and within international relations. The main propositions of this case study are fivefold.

First, British arms exports are not primarily driven by the structure of the defence manufacturing base, but from a more general desire to export what is produced. Second, Britain is motivated by overlapping reasons of foreign policy – influence and trade – in exporting arms and dual-use technology. The evidence here suggests no crude notions of exchange but rather a general pursuit of good relationships. Third, it is not true to say of Britain that it is no longer able to refuse arms transfers exclusively on political grounds.[12]

Fourth, this case study shows that civil and military trade are interrelated in that the willingness to supply defence goods becomes the price of access to the wider civil market: reverse influence[13] obtains.

Fifth, outside of international control regimes, there is no such thing as British arms export policy. This is partly because, as Franklin argued, governments 'are basically responding to requests' from other governments:[14] like much foreign policy, arms export 'policy' is reactive. Competing interests are brought to bear upon each decision. Hard and fast rules about what interests are paramount are difficult to discern given the varying circumstances of each potential export; but foreign policy, as both economic and physical security with the attendant concern for international relationships, remains the strongest determinant. And yet, such decisions are never straightforward: diplomatic, strategic and economic interests have to be tested against each other.

These inductions are located in the general literature, with some preliminary comments about the ambiguity of the previous evidence and the methodology of past studies of the arms trade. To what extent these propositions apply to other suppliers is discussed in the Conclusion.

*1. British arms exports are not primarily driven by the structure of the defence manufacturing base, but from a more general desire to export what is produced.*

The bi-polarity of the post-war era meant that the arms trade was explicable by reference to alliance systems, ideological orientations and bloc structures. The domination of the arms trade system by the two superpowers mirrored their domination elsewhere and was, to state the obvious, a function of their resources. Similarly, in the multi-polar inter-war years, the arms trade reflected the international system: in the absence of ideological stimuli and restraints, it was unregulated and driven by commercial considerations.[15] These characteristics can also be seen in the post-bi-polar arms trade system: it is, as Harkavy put it, 'de-politicised', its central dynamic is commercialism, as, simultaneously, Third World countries have freed themselves from dependence upon superpower supply, and sellers, driven by economic pressures to sell arms with less discrimination, have multiplied.[16]

In the Cold War period, most theorists tended to see a dichotomy between the motives of superpowers and those of lesser nations. Krause summarized the consensus:

> the motivation of second-tier producers for participating in the arms transfer system is distinct from that of first-tier suppliers: low levels of domestic procurement and relatively small research and development establishments force a greater reliance on exports and forswearing of the potential political benefits that can be derived from arms transfer relationships. The shift in British and French policy in the 1960s to a more commercial orientation (and the rapid rise in their exports in the 1970s) and West Germany's emergence as a major exporter in the face of serious domestic political inhibitions confirm this phenomenon.[17]

It was commonly argued that, whereas arms exports were of economic value to the US and the former Soviet Union, for secondary suppliers, exports were, and continue to be, a *necessity* in preserving strategically important production lines. Dependence upon arms sales is demonstrated by an analysis of the percentage of production that is exported, although no analyst seems ready to fix the exact percentage point whereby states become dependent on weapons exports, thus demonstrating one of the major weaknesses of statistical analysis. For example, Catrina noted that 'the United Kingdom was more export-dependent than the major two suppliers, but clearly less than France'.[18] Britain exports approximately 30 per cent of its military production, France some 40 per cent.[19]

The apparent exigencies of the British defence manufacturing base are reflected in a 'disposition to sell':[20] the UK has pursued a number of tactics to increase its exports of defence equipment. These include: a willingness to facilitate offset deals;[21] the use of aid in underwriting arms sales;[22] and the provision of 'a subsidy for sales to developing countries'[23] in the form of disproportionate amounts of credit for military, as opposed to civil, exports. More general government assistance is provided to arms exporters, through the Defence Export Services Organisation (DESO), in terms of advice on markets and contracts, and the facilitation of access to overseas decision-makers. Assistance is also given in negotiating defence deals and administering them through, for example, inter-governmental agreements or memoranda of understanding.

While such efforts on behalf of military, as opposed to civil, industry are seen by analysts as evidence of Britain's export dependence, the National Audit Office (NAO) argued that:

> Because of the nature of both the products marketed and customer requirements, the government must visibly support the export efforts of the UK defence industry.[24]

Moreover, civil exports are supported by the DTI, Ambassadors and by Commercial Attachés, the latter, unlike their military counterparts, spending all of their time promoting trade.

The macro-economic benefits of arms exports include the contribution of defence sales to the balance of payments, and the maintenance of employment. The principal micro-economic benefit of arms exporting is seen to lie in reducing domestic procurement costs. These benefits are contested, not least because of the absence of reliable data. However, analysts also argue that such benefits are not nationally significant: for example, arms exports contribute about 2 per cent to Britain's balance of payments and provide some 150,000 jobs.[25]

However, this study shows that decision-makers do not share the perception that the general economic benefits of defence sales are insignificant. If defence exports are located in the economic context in which Britain finds itself, rather than simply linked to the problematical nature of the defence manufacturing base, a different explanation suggests itself. The motive is economic rather than strategic. Britain has a long mercantilist tradition; however, it has lost export share from more than 20 per cent of the world total in 1953 to less than 8 per cent in 1990. Moreover, on the whole, it tends to export 'relatively low-quality, low-technology goods' and, since 1983, is a net importer of manufactured products.[26] Arms sales, therefore, buck the trends of

Britain's economic decline. As Alan Clark said in his *Diaries*, 'Weapons are our best exports, and unlike most of the manufacturing sector there are still some bright young people around';[27] or, as Franklin noted, arms represent an area of industrial capacity where Britain enjoys certain competitive advantages.[28] More generally, as another former Minister of State for Defence Procurement argued, 'It follows automatically that if you are good at the job, you ought to be selling your products overseas'.[29] Arms exporting stems from the general economic system, a conclusion implicit in Vayrynen's analysis of socialist and Western arms suppliers. He found political motives operating in the former and economic ones in the latter.[30]

*2. Britain is motivated by overlapping reasons of foreign policy – influence and trade – in exporting arms and dual-use technology.*

SIPRI argued that evidence of the UK's lack of political motives was to be found in its ceasing to provide arms as gifts.[31] However, the UK was, as Brzoska and Ohlson pointed out, the first West European country to follow the US lead in commercializing arms exports in the 1960s.[32] Both Pierre and Krause[33] noted the changing declared aims of British arms export policy. Pierre noted that, in 1955, the Government couched its arms exporting goals in strategic terms:

> the general policy of Her Majesty's Government on the sale of arms is primarily governed by political and strategic considerations: only when these have been satisfied are economic considerations – i.e., the contribution of arms sales to export earnings – taken into account.[34]

By the 1970s, arms exports became framed in economic, not political, terms. However, as Freedman noted:

> By stressing the commercial benefits, the Government helps to confine the debate, leaving the critics either contrasting economic expediency with some higher morality, or else attempting to solve the unemployment problems that would be created by a withdrawal from the arms trade.[35]

Pierre spoke of Britain as lacking the 'strategic rationales' for arms exports, such as securing basing rights. Now that the Empire has gone, and 'Britain's armed forces (are) now almost completely based in Europe', Britain has, he argued, no motives of power driving its arms export policy.[36] However, if we take a wider view, and one that is less crude in its notion of exchange between countries, it is possible to see that arms exports can serve foreign policy goals for secondary suppliers. It is more appropriate to see arms exports as merely part of the web of interactions between countries – a web which supports good relations and all that that brings in terms of influence, diplomatic support, trade and so on.

Indeed, the importance of arms exports as foreign policy tools is enhanced for secondary suppliers in that, in creating, fostering and maintaining relationships, defence sales represent one of the few instruments readily available since second-tier nations are less able to provide alliances or other forms of military, and civil, aid. This sort of benefit – the nurturing of good relations – is not, in contrast to the granting of basing rights, for example, measurable or observable. Because of this, it is often mentioned in the literature as an effect of, or as a motive for, arms exports, but it is rarely explored.

Outside the contexts of international control agreements, scholars have found it difficult to locate foreign policy motives in the sale and non-sale of British military equipment. This might be, in part, because it is difficult to define British foreign policy itself beyond its setting in the northern hemisphere. Alternatively, the failure to distinguish foreign policy motives behind British arms exports is because the literature defines foreign policy too crudely. With the overwhelming emphasis in the arms trade literature upon state-centric realism against the backdrop of the Cold War, foreign policy is defined in terms of power.

While power analysis might be an essential tool in explaining outcomes in international relations, it is less useful in seeking the motives which inspire action since these are a mixture of both realism and aspiration. Foreign policy-making takes place, as Brecher *et al* have argued, in both operational and psychological contexts.[37]

This psychological, or 'aspirational'[38] dimension is largely absent from analysis of the motives of arms suppliers. Yet it is fundamental, for as Boulding argued:

A decision involves the selection of the most preferred position in a contemplated field of choice. Both the field of choice and the ordering of this field by which the preferred position is identified lie in the image of the decision-maker.[39]

The fact that the large body of theoretical work on belief systems[40] has not found its way into the arms trade literature is perhaps not surprising, both because of the lack of data about decision-making processes, and a largely quantitative approach. But the work on belief systems is especially pertinent to Britain as a supplier of arms. Whereas, in the operational environment, it might be broadly true to say, as Morgenthau did, that 'statesmen think and act in terms of interest defined as power',[41] in Britain's psychological milieu, as Christopher Hill noted, 'Nostalgia . . . has continually affected the formulation of modern British foreign policy.'[42]

It is the influence of the past, both psychologically and operationally, that makes British foreign policy so difficult to define. True, one can distinguish certain policies with regard to the US, Europe and the Commonwealth – Churchill's concentric circles – although these fluctuate with time, governments and personalities. But British foreign policy cannot be encapsulated in a simple mission statement in the way that 'containment' summarized American post-war foreign policy.

Attempts to deal with current realities overlay historical legacies. The past, as Christopher Hill put it, 'is more than an influential memory; it is seamlessly stitched in to the present'.[43] Whether this past gives Britain, as Hill asserted, an influence 'beyond its ranking on economic indicators'[44] or whether it results, as David Sanders argued, in Britain simply 'maintaining an audible – but almost invariably ineffectual – voice in international diplomacy'[45] is beyond the scope of this book. What is important to note, however, is that, as Lord Franks said, 'it is part of the habit and furniture of our minds that Britain should be a great power'.[46] This was evident even in the 1980s. As Lady Thatcher argued:

> We developed what might be called the 'Suez syndrome': having previously exaggerated our power, we now exaggerated our impotence. Military and diplomatic successors . . . were either dismissed as trivial or ignored altogether. Defeats, which in reality were the results of avoidable misjudgement, such as the retreat from the Gulf in 1970, were held to be the inevitable consequences of British decline.[47]

The quotation illustrates the conflict in British foreign policy between objective constraints and human perceptions. The UK might well be limited by its relative power but that does not prevent a behaviour unlimited in its ambitions. With regard to arms transfers, there is no reason to suppose that these great power pretensions do not operate in the directions of both constraints and sales. In the latter, Britain seeks to use arms sales as a means of maintaining links with former colonies or simply seeking influence wherever it is to be had: arms supply, and non-supply, consciously or unconsciously, extend the 400-year-old tradition, in the absence of aircraft carriers, of Britain's global reach. Or, as a senior MOD official put it, 'arms sales are an expression of foreign policy'.[48]

There is a perception among British decision-makers that military exports express foreign policy in a number of ways. Both Alan Clark and Sir Adam Butler argued that Britain gained influence through arms sales. The former said that they 'stitch you into the society' by establishing links with the armed forces, who, in any given Third World country, tend to be one of the most influential organizations.[49] A DESO official argued that 'real influence' came from arms sales if the potential recipient could not acquire the desired equipment elsewhere.[50] In other words, there is a distinction to be made between what Clark was referring to – influence – and what the DESO official meant – leverage. The latter can be thought of as a more precise phenomenon, encapsulating that crude notion of exchange which most arms trade analysts look for in measuring the political effectiveness of arms exports. Weapons sales can be seen as part of a more general economic interchange between states; and it is 'economic activity (which) forms a much more practical basis for improving relations'.[51]

The other, overlapping, foreign policy dimension of defence sales is trade. Again, this is part of Britain's past, but it is also part of the present. Sir Adam Butler saw commercial concerns 'as a common strand running back for a long time through our history', and British foreign policy goals as being primarily concerned with trade.[52] Alan Clark concurred with this view. Outside of the East–West conflict, he said, 'foreign policy is about markets'.[53] Arms transfers help secure the *general* markets of developing countries.

*3. Britain does refuse arms transfers exclusively on political grounds.*

In part, the ascription of significance to the defence manufacturing base is the outcome of employing, in the absence of data, the rational actor model. Analysts work backwards from outcome (exports) to motive (the perceived importance of the defence manufacturing base). However, it also stems from the methodology of studying arms production and arms exports together, and from isolating conventional arms from non-conventional ones. Brzoska argued that it is 'pointless' to analyze the weapons trade without reference to defence manufacturing.[54] But, while arms production is a necessary condition for arms exporting, it might not be a sufficient one. While the arms trade flows from production, there is nothing automatic about exporting since government stands between the two processes. Selling arms, therefore, does not have to be explicable simply in terms of their manufacture.

The trade in conventional arms is separated, for the purposes of analysis, from the trade in weapons of mass destruction (WMD). At first sight, this seems entirely reasonable: WMD are not traded, save for the United States' export of nuclear missile systems to the UK. However, if, as is maintained, production and trade must be treated together and that, more particularly, the arms trade is the result of the imperative to maintain production capacity, one would want to ask why countries, such as France and Britain, are not similarly compelled to export WMD and ballistic missiles or related technology.

While there is, of course, evidence of the diffusion of weapons of mass destruction, at the same time there are unprecedented, albeit imperfect, efforts to control their spread. The principal control regimes in this respect are: the Nuclear Non-Proliferation Treaty (NPT) 1968; the Geneva Protocol on Chemical Weapons 1925 (since superseded by the Chemical Weapons Convention (CWC) 1995); and the Biological Weapons and Toxins Weapons Convention (BWC) 1975. The Missile Technology Control Regime (MTCR) 1987 seeks to control the diffusion of the means of delivering WMD over long distances and without warning. Britain is a member of all of these control regimes.

Two control regimes were set up to support the NPT in the 1970s: the Zangger Committee and the Nuclear Suppliers Group (NSG), the latter also known as the London Club. The former produced a 'trigger list' of dual-use materials and equipment which would not be exported if destined for facilities subject to International Atomic Energy Agency (IAEA) safeguards. The NSG extended that trigger list and asked its members to seek assurances about the end use of exported uranium enrichment technologies.

The UK also belongs to the Australia Group, set up in 1985 following the use of chemical weapons in the Iran–Iraq War. This regime established an information exchange about both chemical and biological weapons proliferation, and formulated a list of precursors – the main chemicals used in weapons development – which totalled fifty by 1991. Nine of these chemicals constitute a 'Core List' and are the subject of national controls; the remainder comprise a 'Warning List' which is circulated to industry. The UK had already brought eight of the nine 'core' chemicals under statutory control in 1984. Before Iraq's invasion of Kuwait on 2 August 1990, the UK had brought thirteen of the 'Warning List' precursors under statutory control. It should be noted that the export of these chemicals is not banned, but controlled – i.e. an export licence is required – 'since all but two of them have perfectly legitimate civil uses'.[55] Thus, as is the case with dual-use goods under NSG controls, the question of end use assumes central importance.

In January 1988, a set of 'guidelines for the overseas promotion and supply of nuclear, biological and chemical (NBC) defence equipment and the provision of NBC training' was circulated across the DTI, MOD and FCO. These guidelines directed that sales of defensive NBC items 'should be limited to quantities not exceeding the requirement of (the recipient's) Armed Forces'. Furthermore, such exports were to include a 'no transfer clause' in the contracts 'through which the purchaser undertakes not to re-export or to use the equipment except for the purposes of self-defence'. A démarche is issued to this effect before export can take place.[56]

The MTCR constitutes a set of guidelines on the transfer of missiles, missile sub-systems and associated technology. According to the Treaty, 'the decision to transfer remains the sole and sovereign judgement of the (exporting) Government'.[57] Again, given the dual-use nature of much of the goods in question, the issue of end use is crucial in the export licensing process.

In 1991, the UK published a list of thirty-three 'Countries of Concern' in terms of the proliferation of weapons in the four categories just discussed.[58] Before 1991, however, lists of countries believed to be procuring WMD and ballistic missiles, as well as countries known to be diverting the relevant technology to proliferators, were used for the purposes of internal decision-making. The Government also published, for circulation in industry, and somewhat clandestinely through *Jane's Information Services*, details of non-proliferation controls and countries believed to be engaged in the production of WMD.[59]

Krause noted that 'arms have only rarely and briefly been traded as freely as any other commodity'.[60] This is the point: the export of arms involves the consideration of national security and hence export controls operate for some technologies and for some countries. If we allow that conventional arms exports are of importance to national security in maintaining the defence manufacturing base, there is still a calculation to be performed by decision-makers which involves many, often competing, factors because national security comes in different guises. It is not axiomatic that production results in trade.

A number of other criticisms can be made of the methodology employed by analysts in reaching the conclusion that exports are driven by the need to maintain capacity in the defence manufacturing base. Two criticisms apply particularly to those who adopt a systems approach to the arms trade: a tendency to overlook agency by governments; and an emphasis upon what arms are actually sold.

The work of both Harkavy and Krause, in searching for theories which explain the arms trade as a whole, emphasised structural imperatives. This is a function of the stress upon the arms trade *qua* system: states become units behaving as the environment dictates. Even if it were the case that structural conditions *forced* a decision between arms exports and arms imports, the point is that this might not constitute the grounds upon which decision-makers make choices. Without access to the processes of policy formulation, one cannot know if governments make the same connections, and are as rational, as the analysts.

The second characteristic of systemic analysis is that it tends only to examine what *is* transferred. As Edmonds rightly pointed out, this is an issue which affects all students of the arms trade:

> Only where the government has declared its intention to embargo sales of particular weapons, or prohibit sales to particular customers . . . is the necessary data going to emerge.[61]

The opportunity to study refusals as well as approvals of arms transfers enables one to see the exercise of choice by government. Just as the British Government chooses not to export WMD and ballistic missiles, so too one sees it choosing not to export some conventional weapons to some countries.

The stress upon the economic motives of second-tier suppliers means that there is a tendency for researchers to present the arms trade as 'de-politicized'.[62] However, the export of arms can never be a purely commercial act. It is inherently political, not just because of its intimate connection with sovereignty and with the diffusion of military capabilities abroad, but because the export (or non-export) of arms is viewed as a political statement, not only by the potential recipient but by all other states connected to both seller and purchaser. Arms, in Luckham's phrase, are 'a political commodity'.[63] Or, as Pierre put it, weapons exports 'are *foreign policy writ large*'.[64]

This recognition of the intrinsic political significance of arms transfers is widespread in the literature but it is never connected to the motives of non-superpower suppliers. Because arms supply, or a refusal to supply, has political meaning to the recipient, it necessarily follows that there are political costs and benefits associated with the decisions of suppliers. These costs and benefits are not explored in the literature in terms of the foreign policy uses of arms by secondary suppliers.

Furthermore, it was not just the superpowers who refused arms sales purely on political grounds because East–West conflict also affected the policies of second-tier suppliers. In the Western case, this operated formally through the mechanism of the Co-ordinating Committee on Multilateral Export Controls (COCOM), and more importantly, politically, by virtue of ideological disposition. Apart from the MTCR, COCOM represented the only internationally agreed restraint upon the transfer of conventional weapons and technology (excluding UN-sponsored agreements on some so-called 'inhumane weapons'). COCOM was created in 1949 and operated throughout the Cold War as a means of preventing strategically important technologies from reaching the Warsaw Pact countries as well as Mongolia, North Korea, Vietnam, Albania and China. COCOM had no accompanying treaty but the regime's export criteria were incorporated into each member's national legislation. High-level meetings were sometimes called to determine strategy and, from 1989 onwards, to relax and dismantle the regime. A rolling review of the list of COCOM-controlled goods was conducted. A COCOM Co-ordinating Group processed requests – by time-consuming circulation among member states – for exemptions from the embargo. Equipment lists – Industrial, Military and Atomic Energy – operated in tandem with a list of proscribed, as well as diverter, countries.

Tensions in the COCOM regime were evident during the 1980s. As a senior DTI official put it, 'There were different philosophies of control in COCOM.'[65] Put another way, the US was much more committed to a stringent interpretation of the regime's export criteria than the Europeans and, as the most important member, American strictness automatically set up a conflict, particularly for Britain, between two sets of interests – alliance relations and trade. Douglas Hurd, then Minister of State at the FCO, described the UK's occasional proposals to make defence sales to China as 'a battleground'.[66]

That Britain was 'chaffing' under the COCOM regime throughout the 1980s is evident.[67] This irritation should not be thought of as merely stemming from an impingement upon Britain's economic interests: it can also be seen as having political and administrative dimensions. In respect of the latter, a DTI official remembered: 'World trade was rising, but the list was not being reduced at the same speed that the applications were rising. So we were catching more and more things in the licensing system.'[68]

The 1979 Export Administration Act, strengthened in 1985 to demand that countries seek American permission before transferring goods of US origin, was illustrative of the different approaches to trade with the East on each side of the Atlantic. On the one hand, the clash can be seen as having an economic root, but, on the other, it was symptomatic of differing strategic perspectives: the Europeans had been more willing to engage with the Warsaw Pact, both politically and economically, while the US had favoured, especially in the 1980s, a more confrontational stance.

Nonetheless, for all the tensions within the COCOM regime, for Britain, from 1949 to 1990, the primary objective of the licensing system was 'to stop volume exports to the Soviet bloc'.[69] As such, Britain's export policy was fundamentally political, dictated by the exigencies of the Cold War. Thus, foreign policy was, as it remains, paramount in arms export policy.

Outside of COCOM, Britain abides by arms embargoes, agreed by the United Nations and the European Union, but also refuses requests for sales where its reputation for international responsibility – at home and abroad – will be damaged. Blackwell found, in a survey of British élite attitudes to foreign policy, a belief that 'Britain was the "moral" leader of the world'.[70] The Foreign Secretary in 1988 argued that, because of Britain's permanent membership of the UN Security Council, 'people have looked to us as one of those who ought to play a leading role'.[71] However, such aspirations are not a constant guide to action: harder heads sometimes prevail. Indeed, there is a continuing conflict between perceptions of Britain's world role, played out through arms sales decisions, between pretensions to moral leadership and the necessity for trade. Outside of the contexts of East–West conflict and weapons of mass destruction, one sees a competition between these two strands of foreign policy. Both strands operate as springboards for *and* constraints upon arms sales.

A senior diplomat made the general point that one 'cannot implement guidelines on defence sales without taking account of foreign policy'.[72] Defence export controls are manipulated in support of foreign policy. The Prime Minister argued that this was 'right'.[73] A former Minister of Trade asserted that 'Export controls are a perfectly legitimate instrument of foreign policy';[74] and another Minister of Trade observed that export control replaces the imperial gunboat.[75]

*4. Civil and military trade are interrelated in that the willingness of Britain to supply defence goods becomes the price of access to the wider civil market: reverse influence obtains.*

Weapons exports have a general economic importance in that they are bound up with civil exports. Inevitably, since 'Britain does live by trade',[76] exports of arms became intrinsically important. An official from the DESO argued that the organization was inspired to promote defence exports not out of a precise concern for capacity in the defence manufacturing base, but from a general desire to improve Britain's economic performance.[77] Moreover, in some circumstances, arms sales can act to open up the general market, especially where, as Sir Adam Butler argued, economies are centralized and government is in a position where it buys non-military as well as military goods.[78] But, perhaps more importantly, potential recipients of arms have tied defence exports to the supplier's access to the general civil market. An interview with the Shah of Iran in 1977 is perhaps one of the earliest obvious examples of this:

*Shah:* They (the US) are free to sell their arms to whom they please as I am free to sell my oil . . . Remember that Iran could be a ten billion dollars-a-year market that you (the Americans) would lose.
*Q:* Do you mean that if the US does not sell arms to Iran, Iran would curtail civilian purchases from the US?
*Shah:* Yes, of course.[79]

Repeatedly, during evidence before Sir Richard Scott, officials and Ministers referred to their fear of reprisals against the UK's civil exporters if defence goods were constrained in respect of both Iran and Iraq. Lest this should be thought of as mere exculpation, one finds references to the intimate link between civil and defence exports in the papers surrounding export licensing decisions and policy formulation. An internal DTI memorandum makes the general point: 'Our prospects of making . . . exports depend to a greater or lesser extent on the UK's willingness to supply arms and related material.'[80]

*5. Outside of international control regimes, there is no such thing as British arms export policy.*

Few of the concerns of British foreign policy find themselves expressed publicly in respect of arms export policy. Rather, the general economic benefits of weapons sales are frequently rehearsed by the Government and reference is made to the traditions and responsibilities of a great power in the published criteria for assessing defence exports. In 1981 the FCO, in setting out the foreign policy uses of arms exports, expressed those uses largely in terms of East–West conflict.[81] Because the real sources of foreign policy are not publicly spelled out, and it is only those factors which will constrain rather than inspire sales that are generally revealed, it is easy to see British arms export decision-making as 'reactive'.[82]

An examination of the national guidelines which govern British defence exports shows that such guidelines constitute a *framework* for the competing interests and subjectivities which surround judgements about export licence applications. The decision-making process cannot be understood unless one first understands that there is no such thing as British conventional arms export policy: defence sales are an expression of *other* policies.

In 1981, in a memorandum to the Foreign Affairs Committee of the House of Commons, the FCO set out the national criteria for the control of overseas arms sales. Controls were imposed upon countries:

> which pose a direct threat to the safety of Britain or our NATO allies; . . . (which are) covered by a mandatory UN embargo; (and) to regimes to which special political considerations apply, (for example, Taiwan and North Korea).

The FCO memorandum went on to say that, in other cases, 'the basis for assessment' would include: the stability of a region; the safeguarding of British dependent territories; multilateral arms control (for example with regard to 'inhumane' weapons); 'the interests and attitudes' of allies and other friendly countries; and the nature and potential uses of the equipment. On this last point, the FCO argued that two distinctions could sometimes be made: one between offensive and defensive equipment; and the other between items which could be used for 'internal repression' and items which could not be so used, for example, naval patrol craft.[83] These factors are identical to those in a set of internal FCO guidelines, except that the latter also include a consideration of whether 'terrorist attractive equipment' would be released to certain countries.[84]

All of the criteria are capable of differing interpretations. The criterion of 'security' ought to be the least ambiguous, yet, of course, as a largely subjective notion, it is not. As Lt Col. Glazebrook of one of the MOD's security branches put it, 'Every sale is a loss of security.'[85] It is this room for interpretation which necessarily invites dispute into the decision-making process. Often the differences are about short- and long-term notions of security and the degree of risk involved in any one decision. Arguments about security are complicated when they revolve around technical questions about the level of performance of equipment and, in particular, about their end use. The latter is rarely certain so that all sides can bring their own hopes and fears to the discussion.

Similarly, there are no hard and fast rules which identify arms transfers that might destabilize a regional balance. While the FCO maintains that it was one of its criteria, the MOD did not consider it as a *separate* criterion 'principally because instability is likely to damage commercial interest and our national security'.[86]

The guideline which refers to sales of military equipment to repressive regimes is neither a ban on certain arms, nor is it an abstention from the defence trade with certain countries. As a senior FCO official put it: 'If we were to trade only with those countries with an unblemished human rights record, we would end up trading only with Botswana.'[87]

As for equipment, in 1989 Alan Clark argued: 'It is impractical to seek to control materials capable of being used for torture since almost any material could be put to use.'[88]

An Assistant Under-Secretary, Commitments, in the MOD confirmed the fact that this guideline springs from a concern for national reputation and is thus applied unevenly, depending upon the circumstances. An internal memorandum from June 1990 argued that:

Ministers will always have in mind political and presentational factors, for example in relation to particularly odious regimes. These factors may from time to time rule out exports that may otherwise satisfy the criteria (of commercial interest and national security).[89]

A former Minister further noted that:

There is a high degree of hypocrisy which exists throughout the world of arms sales and . . . we are not totally free of that. Are you totally pure when you see the French selling more strongly than anybody, regardless of what the human rights issues are?[90]

In a similar vein, the morality of selling arms to states in conflict is subordinated to commercial interest and national security. Moreover, the international law of neutrality, which carries certain obligations about the supply, or rather non-supply, of military assistance, is interpreted by Britain, and other countries, with a degree of flexibility which would be seen as disregard in other circles.

All European countries, according to a survey by the Presidency of the European Council, profess to applying very similar constraints to those discussed here:[91] indeed, the European Union agreed eight Criteria on Conventional Arms Transfers in 1991 and 1992. But constraints are expressed as guidelines rather than as a list of equipment that cannot be exported or as a list of countries to which certain goods cannot be sold. And they are expressed as guidelines so that national interests might be brought to bear on decision-making. Thus, in spite of occasional British professions of willingness to abide by multilateral conventional arms control, as well as an apparent enthusiasm for the UN Arms Register:

> Generally speaking . . . it can be said that arms sales are likely to be restrained, if at all, by practical calculations of national foreign policy rather than by general principles.[92]

To underline the point, the Head of the DESS in the MOD privately noted that: 'Usually, in fact, defence sales are governed by international obligations, commercial interests and national security.'[93]

The mechanism that allows all of these considerations to come into play is the export licensing system. It is here that it becomes apparent that there is an absence of arms export policy *per se*: what there is is a legislative framework. The Government is able to exercise control without, as Alan Clark has it, 'proper approval'[94] because of the nature and extent of powers inherent in the legislative framework of export control. The principal legislation was the Import, Export and Customs Powers (Defence) Act 1939, a law introduced to establish export control *vis à vis* the Second World War. The Act was given permanent status in 1990 in the Import and Export Control Act. It empowers the Board of Trade to control any exports from the UK and such controls are given effect by the Export of Goods (Control) Order (EGCO) which lists those items requiring licences. The Order can be amended at any time without Parliamentary procedure and thus with great speed. The three former COCOM lists form the basis of all export control. In 1985, a fourth list was added to cover the export of technological documents relating to 'goods, technologies and processes' not in the public domain.[95]

However, the Government has even greater flexibility than that legislatively available to it. By communicating guidelines to exporters, it can exercise more, or less, constraints than are contained in the EGCO. Goods brought under legislative control for one reason can be refused a licence on other grounds, for example, political circumstances. In effect, the Government can act to control exports both formally and informally.

Three types of licences are issued: permanent, temporary, and trans-shipment. Only in 1989 did it become possible for the DTI to exercise a sanction against companies which did not ensure the return to the UK of goods exported on a temporary licence.[96] Permanent licences are divided into four categories: an ordinary export licence; an open individual export licence (OIEL); an open general export licence (OGEL); and a bulk licence. The OIEL authorizes a company to export specified goods to particular countries up to a certain value and typically covers the supply of spare parts – but not ammunition – as and when the customer requires them. The OGEL differs from the OIEL only in that it can be taken advantage of by any company. Such licences were introduced in 1989 for EC and COCOM partners for all but the most sensitive goods. Bulk licences authorize the export of a quantity of broadly described goods, such as spares, up to a specified value, and are commonly used for the export of ammunition.

The EGCO empowers the Government to revoke a licence at any time. However, the Government does occasionally make *ex gratia* payments as compensation for the consequences of such actions,[97] even though companies can and do incorporate into contracts export licence break, or *force majeure*, clauses to cover this contingency.

Finally, Article Five of the EGCO states that exporters are to provide 'proof' that exported goods reached their authorized destination.[98] However, the £2,000 penalty for failing to provide such proof might or might not constitute a disincentive to the exporter, depending upon the value of the contract. The issue of end use is one beset by some confusion. First, it is necessary to distinguish between 'end user', to which Article Five above refers, and 'end use'.

End user statements represent both the verification of the existence of an end user and an undertaking by the importer not to re-export the goods. Such undertakings are not requested from Government organizations. However, in government-to-government agreements, such as memoranda of understanding, disposals to third parties are prohibited. This is described as a 'standard clause'.[99] Memoranda of understanding on security might also be used to safeguard certain equipment, from either disposal, or exposure, to a third party.[100] The system of end user undertakings is, in effect, one of 'honour';[101] or, to quote Alan Clark, end user undertakings 'are not worth the paper they are written on'.[102] The Government, as Pearson said, 'professes to rely on intelligence information to track down harmful re-exports and the threat of future sales bans to discourage them'.[103]

Except in relation to dual-use goods destined for COCOM-proscribed countries, where the recipient had to undertake not to use the goods for military purposes,[104] end use assurances seem to have been rarely requested by the UK.[105] While end user undertakings also require the recipient to state the use to which equipment will be put, it was only in April 1990 that end use undertakings were requested on a more systematic basis. Officials argued that the neglect of the issue of end use was the result of employing 'the COCOM procedures for our purposes'.[106] Since the focus was upon preventing strategic exports to the Soviet bloc, directly or via third parties, the issue of proliferation – the end use of dual-use goods – was paid less attention. However, since the supergun affair, the onus for ascertaining a legitimate end use has been placed upon the exporter.[107]

As in the case of end user undertakings, the Government cannot enforce end use assurances: it can only punish violators with the refusal of further exportation. But such punishment must necessarily be *ad hoc* since: 'After delivery we do not monitor on a regular basis the use which a purchaser makes of such equipment. It would not be practicable to do so.'[108]

However, it should be noted that the intelligence agencies do monitor both end use and end user, and exporters can be prosecuted if they *knowingly* state incorrectly the precise purpose to which the goods will be put.[109]

Enforcement of the licensing system rests with Customs and Excise working with the DTI. Essentially a limited exercise, enforcement might sometimes prevent goods from reaching the wrong destinations. Customs and Excise can prevent exports only if they have timely intelligence. Primarily, enforcement exists to prevent the export licence from becoming 'a kind of dog licence'.[110] The Head of the Export Control Organisation argued that enforcement lends some credibility to a system which is 'self-regulatory'.[111] However, this comment should be treated with some scepticism. On the one hand, the Government has wide-ranging powers; and, on the other, companies work closely and co-operatively with Whitehall so that informal controls and sanctions operate effectively for the most part. Nor should the DTI's comments obscure the fact that the process of export control is able to fulfil the

Government's varying and wide-ranging interests and intentions in the supply and non-supply of military equipment overseas.

# 2

# Decision-making
# for British arms exports

## Models of decision-making and political responsibility

A thorough examination of the defence exports decision-making process is necessary for several reasons. The nature of the British Executive is a matter of both academic and political debate, the chief issue of which is the location of responsibility for decisions. This, in turn, has two dimensions: the extent to which the centre, Cabinet and Prime Minister, has the ability to direct policy; and the distribution of power between civil servants and Ministers within Departments. These issues are of central importance. If one accepts that power rests as much with civil servants as with Ministers, then the notion that Parliament holds Ministers accountable for the actions of the Executive is automatically undermined.

Sir Richard Scott accepted Sir Robin Butler's distinction between responsibility and accountability:

> Ministerial 'accountability' is a constitutional burden that rests on the shoulders of Ministers and cannot be set aside. It does not necessarily, however, require blame to be accepted by a Minister in whose department some blameworthy error or failure has occurred.

While Sir Richard argued that the 'corollary' of such a definition is the 'obligation to be forthcoming with information about the incident in question',[1] it remains problematical in three respects. First, if Ministers are not responsible, it must be assumed that civil servants take on some constitutional duty, as well as a moral right, to speak, not on behalf of Ministers or Departments, but for themselves. Second, it raises the question of the extent to which Parliament ought to be able to hold Ministers to account. That is, if errors are committed, it could be argued that Ministers remain responsible in that their role is to provide management: the re-definition of Ministerial responsibility removes the ultimate incentive for Ministers to perform their duty. Finally, the sort of information that would ensure that Parliament knew where responsibility lay between Ministers and officials is never likely to be disclosed: informal, unrecorded discussions and instructions will remain hidden.

The events discussed in this book do not support the notion that the principle of ministerial responsibility requires re-definition. The actions of civil servants accorded with the political desires of Ministers: implementation followed policy intentions. The issue of where power lies in the decision-making process is particularly important for this study in that, to explain Britain's defence exports policy towards Iran and Iraq, it is necessary to distinguish policy, the decisions of Ministers, from failures to implement policy, errors or wrong-doing by civil servants. This chapter is also necessary to discover the nature of defence exports policy.

The main body of academic work which has as its central contention that governmental decision-making is characterized by both muddle and an uneven distribution of power comprises the bureaucratic politics and the organizational process models. The former is composed of two basic propositions: within government, power is diffused and interests are a mixture of the personal and the organizational. Thus, any decision is the result of bargaining and, as such, is neither rational nor inspired by a conception of the 'national interest'.[2] That this model has its critiques, even as it applies to the larger, pluralistic government of America is one issue;[3] but the important concern for the purposes of this chapter is its applicability to the British system of administration.

William Wallace argued that the bureaucratic politics model is not useful in the explanation of British external policy:

Whitehall is not Washington; the open conflicts between sections of the administration which characterise bureaucratic politics in America have no exact parallel in Britain.[4]

On the other hand, Steve Smith has pointed out that:

the tradition of secrecy makes it impossible to undertake the kind of in-depth study that is necessary for investigating bureaucratic politics types of accounts. But to accept this does not mean that bureaucratic politics or organisational processes are not at work.[5]

However, access to the decision-making process in this case confirms that Wallace is right, at least in respect of arms export policy. First, one sees that power might be diffused in terms of taking decisions *within* the framework of policy, but well-established chains of command ensure that policy is dictated by the centre to which, in turn, Departments continually refer. Second, shared, rather than conflicting, interests are more prevalent within the British administration. These factors also make models of decision-making which stress the power of officials over information and implementation less relevant in the British context: the system is too sensitive, both in organizational and psychological terms, to the political concerns of the centre.

Alan Clark argued:

> policies are neither determined nor evolved on a simple assessment of National, or even Party, interest. Personal motives – ambition, mischief-making, a view to possible obligations and opportunities in the future, sometimes raw vindictiveness – all came into it.[6]

However, these factors do not impact upon arms export policy for a number of reasons. While decision-makers might well substitute the better-defined personal interest for the national interest, this does not apply in the case of military sales in the way that it might in some other policy areas, or particularly in crises, such as the Falklands War. In any case, one can argue that, since the politician's personal motive will be popular support, in consulting his or her personal interests, he or she is *ipso facto* consulting national values. However, because arms sales decisions are partially isolated from the domestic political context, by Government secrecy and by Parliamentary and public ambivalence, the personal ambitions of policy-makers are unlikely to be furthered in this particular arena. Of course, it might be argued that the relationships which exist between some policy-makers and defence industries, (and those who act as agents thereof), supply an additional and important personal interest. However, it is impossible to disentangle the national from the personal interest in these cases, not least because the 'national interest' is so contestable, but mainly because the two can coincide.

The bureaucratic politics model is not relevant to decision-making about arms exports for two reasons. First, one does not see the necessary politicization of policy. There are no personal political motives to be substituted for a conception of the national interest because arms exports are only controversial for a minority of MPs and voters. Second, even if Departments had different perspectives upon military sales policy, the British political system is so centralized that a resolution of such difference would not represent a bargain or a compromise. In any case, the scope for negotiation is somewhat limited. As Edmonds has argued, 'either arms are sold, or they are not'.[7]

Jordan and Richardson, for example, have argued that it is civil servants and pressure groups who constitute the essence of the decision-making process in a fragmented, or, as the Cabinet Secretary called it, 'federal',[8] system of government.[9] Jordan and Richardson argued that the British political system is divided into 'policy communities' with Cabinet unable to do more than give a general direction. However, this notion overlooks the efficient communications system which operates between the Department on the one hand, and Cabinet and Prime Minister on the other. This communications system comprises a powerful secretariat (the Cabinet Office as well as the Prime Minister's Office), and a structure of committees and well-understood procedures. Even minor policy changes regarding defence sales to Iran and Iraq were put through this system. What makes the system work are the political sensitivities of civil servants and the trust which largely exists between them.

Another objection to the idea of quasi-independent policy communities is that it confuses the implementation of policy with the policy itself. This is not to denigrate the importance of implementation: policy intentions can be fulfilled or frustrated by the process; and the policy might well turn out to be in the details. However, in respect of arms exports, the evidence here suggests that the policy intentions of Cabinet and Prime Minister are fulfilled by the Departments which execute them. Again, this is a function of a centralized political process, shared interests, and co-operative ways of working both within and between Departments.

Sir Robin Butler, the Cabinet Secretary, argued that the execution of defence exports policy towards Iran and Iraq 'was all happening below my eyesight level'.[10] Decisions were taken by officials according to standard operating procedures and, as such, Ministers and senior civil servants can be absolved of responsibility. But this is not strictly the case.

While constitutionally 'Ministers have an overall responsibility for the conduct of his or her Department'[11] to Parliament (although this principle has seen a *de facto* erosion in the last fifteen or so years), administration rests with officials. This is the case for a number of reasons. First, Ministers are so very busy that an official filtering system

has to operate. Even with that process in place, Geoffrey Howe calculated that, as Foreign Secretary, he read twenty tons of papers in the
course of one year in his nightly boxes alone.[12] Decision-making
devolves downwards to junior Ministers and thence to officials. Second, Ministers, of whatever rank, rely upon their officials for advice.
Although Ministers can consult their own political instincts and may
seek advice from outside the system, they are unlikely to do so for
small decisions regarding individual export licence applications
(ELAs). Moreover, official information is necessarily refined to make it
manageable for the Minister. As Lord Trefgarne, a Minister for
Defence Procurement, argued, 'I was dependent upon what (officials)
told me or did not tell me.'[13] Judgements, as the Prime Minister, John
Major, said, 'have to be left to intermediaries ... as to whether the documentation should be brought before Ministers'.[14]

However, one should not see the Minister as some hapless victim of
manipulative civil servants. The filtering system operates to the mutu-
al satisfaction of both parties. As Ian McDonald, a senior official in the
MOD put it:

> Ministers prefer ... to get a recommendation. There are situations,
> however, in which civil servants can, and do, put a submission up to
> Ministers, in which they say: 'On the one hand this, on the other
> hand, that.' It is not easy to come to a decision, except on a subjec-
> tive basis. Oh Minister, will you please make up your subjective
> mind.[15]

As the quotation above makes clear, Ministers do have the power of
decision and are kept informed of even some of the smallest decisions,
or, indeed, asked to take them. The amount of paperwork about which
Ministers complain is testament to this fact. The Inter-Departmental
Committee on Defence Sales to Iran and Iraq (IDC) had to refer its *recommendations* to Ministers for their approval.

The political trick which a Minister must perform is to catch those
issues and decisions which appear routine but which are, in fact, of a
high political order in embryonic form. As a former Minister argued:

> So much is going on in Whitehall and Westminster all the time, that
> it is not always inevitably clear that if something arose that was ...
> no bigger than a man's fist on the horizon, that it is actually in a day
> or two going to very big indeed.[16]

But this is an issue of political management rather than one of the power of the civil service. The British policy process is a responsive one. Ministers bring their own perceptions of the Government's policy to their Departments; and civil servants, in turn, are 'aware of the Minister's pre-dispositions because that is part of (their) sensitivity'.[17] At the same time, a Minister who does not appear to follow the Government line will quickly find the Permanent Secretary taking the matter up with either the Minister or the Cabinet Office.[18]

Thus, although some decision-making is devolved and political attention will not be everywhere, a sense of direction radiates from the centre. In the case of arms exports to Iran and Iraq, there was not a single example of a decision taken by an official which did not cohere with the policy of the centre. Thus, what follows is an investigation of the process of implementation which demonstrates this centralization as well as the shared interests of the organizations involved and their largely co-operative ways of working. This is not to say that there are not divergences and conflicts, but those shared interests – the policies of the centre – prevail.

The fact that, in contrast to the US, the British political process is characterized by 'established ties of kinship and culture',[19] as Heclo and Wildavsky put it, makes the exact location of the essence of decision difficult to pinpoint. Much of the process is simply not recorded since 'formal discussion follows after informal chats'.[20] Officials 'clear things in an informal way' because they 'trust one another enough not to have to send bits of paper back and forth'.[21] Government documents are also sparsely written because of the fear of leaks: 'minutes can . . . soften the blows'.[22] Furthermore, civil servants tend to write in, what a senior DTI official called 'rounded terms'.[23] There are no hard and fast rules about the recording of Minister's morning (prayer) meetings because of their informality,[24] but Cabinet Office meetings are not minuted.[25] These shortcomings in the documentation should be borne in mind in this and subsequent chapters.

## Parliament, public and the interests of policy-makers

The role of Parliament and public in arms export policy demonstrates that there is no incentive for policy-makers to substitute personal political motives for national ones, because arms sales do not create the 'mortal danger'[26] for British politicians that other subjects can and do. The reason is primarily a constitutional one: as in all policy-making, Parliament is marginalized. Whereas Congress enjoys the power of the veto, Parliament is unable even to perform effectively its functions of oversight and scrutiny. At a broad level this can be attributed to three causes: legislature and executive are not separated; Parliamentary machinery is inadequate; and party discipline holds sway. At the particular level of military exports policy, it is because of an excess of secrecy on the Government's side and a general deficit of interest on Parliament's part.

The Government will not provide for Parliament the details of sales, nor will it disclose details of the operation, or sometimes the very existence, of some sets of guidelines governing arms sales to particular countries. Baroness Thatcher understood 'that it was rare for such guidelines to be made public';[27] or, as Lord Trefgarne put it, 'Policy is not always everything that is announced.'[28]

Parliament has two principal mechanisms for discovering policy: Parliamentary Questions (PQ); and investigation by Committee. There are two forms of the PQ. Written questions allow Ministers, or rather their civil servants, at least forty-eight hours to respond and, since the questioner cannot ask a supplementary question, they provide a limited means of extracting information. Given this property, the Government tends to treat these PQs with less respect than oral ones: as Sir Stephen Egerton of the FCO argued, 'Interest is shown by tabling an oral question.'[29] Oral questions are seen as combative and the response is therefore 'something of an art form, rather than a means of communication'.[30] According to 'Questions of Procedure for Ministers', there is 'a duty to give Parliament . . . as full information as possible about the policy decisions and actions of the Government and not to deceive or mislead Parliament'.[31] That framework however, does not necessitate 'total disclosure'[32]: 'half a picture' will suffice, provided it is 'accurate'.[33] Timothy Renton, a former Minister of State in the FCO, is worth quoting at length on this point:

> If someone asked me: 'Are you aware of any precursor chemical weapons that have been exported to Iraq?', I would have answered – and I think there is one along those lines – 'No, I am not aware.' I would not have added information about the hydrogen fluoride . . . which went to Egypt, where we were told by one other country that it might end up in Iraq. I would not put that in the (answer), because I had not been asked that.[34]

Public disclosure can also be avoided by the Government taking the questioner into its confidence.

The Committee structure has also proved inadequate in the oversight of arms exports. Again, the official guidance for civil servants giving evidence to Select Committees states that they must ensure the accuracy of that evidence, but, at the same time, they are instructed to withhold information in the interests of the somewhat open-ended concepts of 'national security and/or good Government'.[35] Committees can only secure information via Ministers and it is the latter who decide what information will be given and which civil servants will give evidence, even if a Committee requests the presence of a named official.[36] Thus, in the case of the Trade and Industry Committee's investigation of the export of parts of a supergun to Iraq, the MOD declined to allow two civil servants with knowledge of the matter to give evidence on the grounds that both men had retired.[37] Nor did the intelligence agencies give evidence, thereby inhibiting the Committee's ability to discover the truth since it was their knowledge that lay at the heart of the matter.

While it might be the case that Committees of the House of Commons, like the questions of Members of Parliament, 'are treated with enormous respect and care within Whitehall',[38] their powers in scrutinizing the Executive are circumscribed. In the case of arms export policy, with its elements of commercial confidentiality, large amounts of money, foreign governments, corruption and, for good measure, the frequent involvement of the intelligence agencies, the inherent weaknesses in the system of British political accountability are exacerbated. However, one should also see that, since arms export policy is intimately bound up with foreign policy, accountability will always be curtailed by the discretion with which the British Government perceives it must deal with other countries.

One should note that Parliament, for its part, displays a distinct lack of interest in the subject of defence exports, the 'arms-to-Iraq' case notwithstanding. Edmonds found that, in an eight-year period, only thirteen MPs 'emerged as having demonstrated a continuing interest in one aspect or another of arms sales', with only 261 PQs tabled in that period.[39] In order to wrest information from the Executive, MPs need persistence in following up answers to PQs. In the case of policy towards Iran and Iraq in the period in which they were at war, this was rarely the case. Again, the Trade and Industry Select Committee investigating the supergun affair was limited because the Conservative Chairman, Kenneth Warren, used his casting vote to veto inquiry into the role of the intelligence services and even arranged to meet important witnesses before they gave evidence.

Thus arms exports, in a dichotomous yet disciplined legislature, are neither the stuff by which governments fall, nor a yardstick by which they are ultimately judged. In addition to the facts of secrecy and lack of interest, there are two further reasons why personal political ambitions are neither checked nor furthered in this policy arena. First, in spite of the Labour Party's promises in 1985, and again in 1996, to constrain the selling of weapons abroad,[40] there exists a cross-party consensus about the utility of British defence exports. If arms sales are seen by politicians as promoting the well-being of the defence manufacturing base or as an essential tool of an active foreign policy, then the Labour Party has been as willing as the Conservatives to support these facets of independence. Furthermore, the Labour Party has also been as eager as the Conservatives to grasp an apparent economic success in the midst of decline.

The second reason for the non-controversial nature of arms export policy is the perception that the public is ambivalent about it; at once suspicious of the motives of the players involved, and yet pleased when new orders appear to secure employment. Thus, on the one hand, whatever accusations Parliament might lay at the door of the Executive, politicians:

> measure these things at two levels. They measure them immediately (in) the uncomfortable atmosphere in the House of Commons, but the ultimate decision tends to be in electoral terms.[41]

To state the obvious, governments do not come and go on the basis of their defence exports policy; and, in any case, there is a tendency to view opposition to the sale of arms as coming from 'an unrepresentative part of the public'.[42]

From the evidence in the case of defence sales to Iran and Iraq, there does exist, to use Pearson's word, a 'watchfulness',[43] within government about the public reaction to particular sales. This sense of caution is more noticeable in Britain because, unlike France, the US or Germany, well-organized opposition does exist, primarily the Campaign Against Arms Trade, but also Saferworld, the World Development Movement and the National Peace Council. However, even such well-organized opposition to Britain's participation in the arms trade does not, on the evidence thus far, serve to constrain sales. Rather, in its watchfulness, the Government seeks only to prepare its defences for the inevitable exposures.

## The defence companies

It can taken as a constant that the perspective of the companies will be pro-sales, although they, like the Government, will not want to alienate important customers by selling to the customers' potential foes. But this factor is the only check upon their desire to export whenever and to whatever country has the means to purchase, notwithstanding, as the NAO noted, the apathy of some companies towards exporting.[44]

It is difficult to unravel a company's interest and the state's interest when arms are sold, and there are, of course, examples of the Government constraining arms transfers and thereby damaging some industrial concerns. The influence of companies is more keenly felt at the level of implementation rather than policy-making. Access to decision-makers is built into the process. Ministers, senior officials and the senior management of the arms industries met regularly at the National Defence Industries Council.[45] Again, at a general level, companies gain access via trade associations such as the Defence Manufacturers' Association which now represents the prime contractors as well as the smaller sub-contractors and components makers. The Association attempts to 'ease government regulations and procurement requirements, (and) advises firms on such regulations'.[46] Such associations are, according to Sir Stephen Egerton, 'extremely active and well-informed'.[47] Companies can use such bodies to lobby on their behalf, especially if it is an issue, such as an export licence application, which has ramifications for them all. For example, the Machine Tools Trade Association (MTTA) was able to meet, with great speed, Ministers and officials when there was uncertainty about the export licences of two of their member companies. The larger companies can, of course, individually lobby government.

Companies can also ensure they get the Minister's ear by either complaining to him or her directly or asking the appropriate MP to raise an issue of export licensing with the DTI. Such complaints become 'Ministers' cases' thus ensuring attention by being passed to the Policy Unit of the DTI for action.[48] More generally, defence industries enjoy regular contact with decision-makers; as Lord Howe, the former Foreign Secretary, said, 'There is a constant dialogue between Departments and exporters.'[49] Rob Young, a Foreign Office Desk Officer, recalled:

> quite regular contact with British Aerospace, Westland, Marconi and Racal . . . quite legitimate lobbying on their part when an (export licence application) went in . . . at most levels we find that

desk officers might have contact in a company; or rather the other way round, the company might have a contact at desk level. It was a very frequent and normal part of daily life . . . It was usually very helpful, if at times a little oppressive.[50]

Moreover, defence manufacturers have an inbuilt champion of their interests in the DESO which not only eases the companies' access to Ministers but also shares their perspectives. However, it should be noted that, as the above quotation indicates, it is not simply the DESO which sees a confluence between its and the exporters' interests. To quote the above FCO Desk Officer again:

We had enormous contracts at stake . . . and in those circumstances, the companies relied very heavily on Foreign Office expertise for helping them to sell their goods.[51] (emphasis added)

For those within government who were sometimes less enthusiastic about certain exports, (usually the security and operational staff of the MOD), companies, alone or in tandem with the DTI and the DESO, pursued a number of tactics, not least of which was to argue the business and employment consequences which would flow from a refusal of a licence application. Matrix Churchill used this tactic on a number of occasions. Also used was the inflation of the value of the proposed export 'in the hope that they will persuade Ministers thereby to let it go through'.[52]

In spite of the privatization programme of the 1980s, therefore, the defence manufacturers remain close to the locus of decision. However, there is a price to be paid: 'In return – and this . . . is a completely unwritten bargain – . . . industry accepts that on certain occasions, the Government restricts exports.'[53]

It is usually the defence, as opposed to civil, manufacturers who take the brunt of efforts by the Government to use the export licensing system for the purposes of foreign policy since it is those goods, to state the obvious, that the Government can readily restrict. Thus, the companies accept restrictions in return for export support: the competition between promotion and restraint is at the margins over individual licences; and the relationship between companies and government is a close, co-operative one.

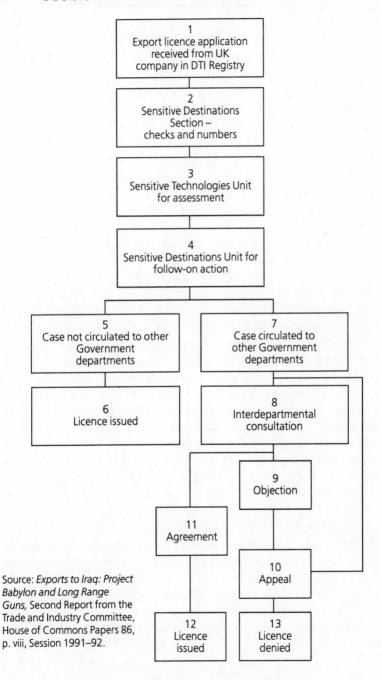

Source: *Exports to Iraq: Project Babylon and Long Range Guns,* Second Report from the Trade and Industry Committee, House of Commons Papers 86, p. viii, Session 1991–92.

Figure 1: Export licence applications procedures

## Inter-departmental decision-making

The diagram (Figure 1) on page 33 shows the inter-departmental deci-sion-making process, although it operates informally through desk officers communicating with one another, as well as formally through committees, principally the Strategic Exports Working Party (SXWP) and the Arms Working Party (AWP). Disputes are resolved through either the Release of Military Information Policy Committee (RMIPC) or through *ad hoc* meetings of either officials or Ministers. The infor-mal veto which each Department enjoys means that difficult decisions will always be transmitted upwards, if necessary to Cabinet and Prime Minister who will, in any case, always take decisions about major arms sales.

The SXWP reports to the Cabinet Office and deals with matters of policy relating to clearance of exports of strategic significance. The Committee is chaired by the Head of the DESS, but has representatives from the three Departments involved in export policy decision-mak-ing. Sub-groups of the SXWP examine matters relating to nuclear proliferation, the MTCR, as well as visas.

The AWP was, to all intents and purposes, a paper committee from 1983 to the early 1990s, operating by correspondence. The AWP gives preliminary clearance to export military goods after thorough exami-nation of the specifications of equipment checked against the company's advertising material. The AWP process also allows the MOD to consider, where necessary, the level of classification – restrict-ed, confidential or secret – of equipment which an exporter wishes to promote. A complex document called 'Table X' is used by the MOD's security branches in making a decision. It is 'key' in that it 'attempts to grade different export destinations by the classifications of equipment which *Whitehall as a whole*, thinks it fit for them to get' (emphasis added). Table X is updated, primarily by the FCO via the embassies, and takes into account such matters as 'the reliability, standards of physical security, (and the) degree of integrity in the public service of the customer concerned'.[54]

The Chair of the AWP is a member of the DESO, but each Depart-ment (DTI, MOD, FCO and the Treasury) has a *de facto* veto. An AWP application is sent by the company directly to the appropriate Region-al Marketing Desk in the DESO which then passes it to the DESS. The DESS acts as the clearing house, seeking the views of the FCO – 'it should always go to the Foreign Office'[55] – as well as the MOD. The form does not go to the DTI, although they 'can make representations'. The DESS also 'attempts to reconcile any differences', and returns the decision to the company via the DESO.[56]

Some 10,000 AWP applications (forms 680s) are received each year and the process takes months to complete as precedence is given to ELAs. However, the system is widely observed by companies since it is in their interests to obtain an early indication of the likely success of a future ELA. An approved AWP application 'is fairly definitive in terms of leading to an export licence unless circumstances have changed dramatically in the intervening period'.[57] Indeed, over a ten-year period, only five AWP approvals did not result in an export licence.[58] It is the AWP system, a good communications network between companies and officials, together with the licensing of many goods for COCOM reasons that are exported to friendly countries, that accounts for the fact that only 200 export licence applications were refused against 97,000 issued in 1987. By the early 1990s, after the introduction of OGELs, ELAS had dropped to 21,000 with about 300 refusals each year.[59]

The last of the inter-departmental committees is the Restricted Enforcement Unit (REU) which brings together the DTI, MOD and FCO with both Customs and Excise and the intelligence services (DIS, the SIS and the Security Service). Unlike the rest of the inter-departmental machinery, the REU exists not to resolve differences, but to overcome the 'rather random way of passing information around'.[60] It works to 'track activity which (is) illegal'.[61]

The REU meets fortnightly in 'swept' premises, is chaired by the Head of the Export Control Organisation (ECO), and is attended by between fifteen and twenty people. Its main weakness, at least in the late 1980s, seems to have been the fact that those who attended its meetings did not pass information gained there to others in their Departments who might have needed it. This was a problem mainly affecting the FCO. Since it was logistically impossible to send all the geographical Desks to the REU meetings, these officials relied upon the representatives from the Science, Energy and Nuclear Department (SEND), the Arms Control and Disarmament Department (ACDD), and the Trade Relations and Export Department (TRED). In respect of Iraq, at least, information was not passed on to the Desk Officer or the Middle East Department.[62] Better lines of communication existed in the MOD and the DTI. A further weakness was that 'a certain degree of rivalry between the participants'[63] persisted.

Because each Department has a veto, each must rely upon powers of information and persuasion if it is to win any particular argument, although the DTI, which has the formal, statutory authority for export licensing, potentially has more leeway than the other Departments in acting independently.

## The Department of Trade and Industry

### Organisation

The statutory authority for the issuing of export licences – the entire system of control – rests with the DTI. A management consultancy exercise in 1987 looked to see if it should remain there and concluded that 'there was an absence of obvious other places to put it'.[64] Sir Richard Scott, in contrast, thought the MOD a more appropriate Department to exercise export control, since 'the effective decision' about defence goods already rests there.[65]

The DTI is also charged with increasing the volume of export trade, although it should be noted that it is the DESO's responsibility to promote military exports. However, civil and military overseas sales are inevitably and inextricably linked. Only in 1992 was the DTI divided, in structural terms, between the two functions of trade promotion and export control.

Trade policy, until 1992, subsumed export control and was the responsibility, with export support, of an Under-Secretary who, in a confusing arrangement, was also responsible as Head of Overseas Trade 2 (OT2) for two geographical branches responsible for the UK's commercial relations. Overseas Trade 4 was responsible for export support in the Middle East and would 'work closely with the export licensing side'.[66]

Export control was located in a sub-division of OT2 called OT2/3 which in turn subsumed the Export Licensing Branch (ELB), the Security Export Control Section (SECS), which provided technical advice about COCOM industrial list goods, as well as a third section promoting trade with China and Hong Kong. In 1988, the ELB became the Export Licensing Unit (ELU) when the ECO was established, bringing into its orbit the SECs, renamed the Sensitive Technologies Unit (STU), as well as a new Policy Unit which was a small section of four or five people set up to 'look at things which were not day-to-day panics'.[67]

### Perspectives

Although the control of exports is the DTI's responsibility, in effect, its wider role – the promotion of trade – conditions the performance of its regulatory function. In short, the DTI seeks to maximize exports. In a ten-year period, there was only one instance of the DTI opposing an ELA when there were no MOD or FCO objections – a ministerial intervention regarding goods for Jordan following Iraq's invasion of Kuwait in 1990.[68] While the promotion of strictly military goods is the province of the DESO, the DTI understands the linkage which some overseas customers make between the purchase of arms and the trade in civil goods. As one official put it, if military goods are refused, 'then quite clearly they might take reprisals against British . . . businessmen and there are precedents where that has been done'.[69] Thus, the DTI has a strong interest in the export of military equipment. The Department is, however, responsible for the promotion of dual-use goods; it is these goods which pose the biggest challenge to the system of control.

While the end use of, say, a tank is obvious and thus the decision to supply or not to supply is a straightforward one, the reverse is the case in respect of equipment which has both civil and military applications. This is a particularly sensitive matter if the items in question have uses in WMD or missile programmes. The DTI sought to err on the side of exports rather than control in such cases. This perspective is in part a function of its wider mission to increase exports, but it is also a function of its under-resourcing.

|  | No. of ELAs received | Ratio to licensing staff |
| --- | --- | --- |
| 1985 | 83,863 | 2,308 |
| 1986 | 89,705 | 2,330 |
| 1987 | 97,842 | 2,477 |
| 1988 | 92,280 | 2,023 |
| 1989 | 75,925 | 1,373 |

Source: Trade and Industry Committee, *Exports to Iraq*, Memorandum from the DTI, December 1990, Memoranda of Evidence, 17 July 1991, Session 1990–91, House of Commons Papers 607, pp. 10–11.

Table 1: Export licence applications and licensing staff

As Table 1 shows, although the ratio of ELAs to licensing staff fell in 1988 following the introduction of the OGEL for the European Community and COCOM partners, even this reduced workload is equivalent, allowing for leave, training and sickness, to about seven applications per staff day. This level of work meant that for 'an awful lot of the time, one was firefighting'. Eric Beston, Head of OT2/3 said, 'We dipped in and out of issues.'[70] Complaints about delays in processing ELAs culminated in 1989 with the Secretary of State becoming 'very angry' and demanding a paper from the Permanent Secretary. Thereafter, the export licensing staff had to produce the monthly report setting out the number of applications dealt with, as well as an analysis of 'Ministers' cases' to see 'whether (the staff) were at fault'.[71] The extent of delays lead to a recommendation in the Scott Report that export licences 'should be deemed to be granted unless refused within a prescribed time after receipt'.[72]

The under-resourcing, and the ensuing pressure upon staff, means that the DTI's mission to export is underpinned by an unwillingness to enforce control on a routine basis. To succeed, argued a DTI official, 'the control system is required to detect only systematic large scale evasion of control'.[73] Not only is the exercise of control thus focused, because of the pressures upon the Department, but the DTI also seeks fewer export controls because: on the one hand, 'the more . . . there are, the easier they are to evade';[74] and, on the other, the system is seen to 'depend upon the co-operation of exporting companies'[75] and the more controls there are, the more that companies are alienated.

The above perspectives, however, are tempered by acceptance, not only that some export controls are necessary for the state's security, but also that export controls will be used as an instrument of foreign policy. The DTI acquiesces in the influence of the FCO over export control because, as the Head of OT2/3 said, 'there are not all that many weapons that a Government can bring to bear to show its displeasure with a particular regime'.

While the DTI 'was uncomfortable about that being used too frequently',[76] DTI officials share with their counterparts in the MOD and the FCO a sensitivity about what is and is not politically acceptable.

Powers
For all that, the DTI seeks to maximize exports and employs a number of tactics in pursuit of that goal. These tactics derive from its statutory authority, its ability to deploy arguments which have political resonance, and its command of information.

The DTI has complete authority to rate equipment: it is neither the responsibility of companies, which only provide the information, nor the MOD. For example, the security branch of the MOD believed that the Jaguar radio was 'designed specifically for military purposes and should always have been Group 1A (military list)'. However, DTI advisers listed it as Group 3F (industrial).[77]

The DTI also issued temporary licences without waiting for the approval of the FCO and the MOD which had long argued that temporary licences should be treated as permanent ones. A former Minister for Trade disagreed:

> Although it would have been preferable to have always consulted the MOD and FCO . . . where there were tight deadlines . . . the DTI would be justified in granting temporary licences unilaterally.[78]

The main problem with temporary licences, from the point of view of control, is that the return of goods cannot be enforced. On a number of occasions, goods sent to both Iran and Iraq for exhibition were refused export licences by the two countries for their return. No action was taken against either country by way of reprisal; nor was action taken against companies which failed to return temporarily exported goods.[79] Moreover, although in the case of COCOM-proscribed or diversionary destinations the exporter had to undertake to keep the equipment under its control,[80] end user certificates were not required.

Some companies, with the knowledge of the DTI, had already sold, or were intending to sell, goods for which they had applied for temporary licenses.[81] For example, the 600 Services Group sold machine tools to the Iraqi MOD from the Baghdad Fair without permanent licences having been issued. Although the Company was 'reprimanded . . . for selling without the DTI knowing the end users, no further action was taken against the company'.[82] As Sir Richard Scott remarked, 'If (the company) is looking for sales, the easiest one to sell is the one that is there.'[83] With collusion from the DTI, the temporary licence is a path around the potential objections to a sale which the FCO and the MOD might make.

This power extends to other licences by virtue of the DTI's statutory authority:

> Although the DTI could, in theory, discount advice from other Departments and issue or refuse licences on the basis of its own assessment of the balance of relative interest, in practice differences of view are *almost always* resolved by inter-departmental discussion.[84] (emphasis added)

This perception of the DTI's authority had two facets. First, as the same senior official quoted above, confessed:

> where it seemed to me that a Department was withholding its agreement for a pretty pedestrian, nitpicking reason, we would . . . issue licences although we had not formally received clearance.[85]

Second, the DTI controls which ELAs it passes to the MOD and FCO for assessment. The MOD automatically receives all Munitions List items and dual-use goods covered by the NSG, MTCR and Australia Group regimes, as well as dual-use items for COCOM-proscribed destinations. A senior MOD official stated that about 10 per cent of all ELAs were sent by the DTI to the MOD,[86] while one-third of ELAs are circulated to both the MOD and the FCO.[87] The fact that more ELAs are sent to the FCO than the MOD indicates the important role which foreign policy has in the export control process.

The MOD and FCO could request that additional categories of ELAs be sent to the Departments for assessment, but this seems to be an area of contention between them and the DTI. A senior MOD official said that discussions on the subject had continued 'over many years'.[88] Similarly, in respect of the FCO, on at least one occasion that can be documented, the appropriate Desk did not receive the ELAs it had requested for precursor chemicals.[89]

Where the DTI cannot act alone, it depends both upon mobilizing arguments for supply and marshalling its control of information in ensuring the approval of the FCO and the MOD. The Department applied pressure on behalf of companies by stressing the employment and business consequences of a refusal of an ELA. Such arguments carried considerable force with both the MOD and the FCO, increasing in weight the greater the value of the ELA under consideration.

Where other Departments wanted to revoke a licence, or refuse one where a contract was already in place, then the DTI asserted that the Government would face claims for compensation. The EGCO gives the Government complete power to revoke licences, but the DTI argued the case anyway.[90] If a licence was refused, the DTI could explain to the exporter the reason for the refusal except where doing so would jeopardize relations with the potential recipient. Asked if the DTI would comment negatively upon decisions and suggest the company take the matter up with the Minister, the Head of OT2/3 replied, 'I have no doubt that it did occur.'[91]

It is in respect of information about the end use or end user where the DTI has greatest power. Although it is the DTI and not the company that is responsible for classifying goods, the company is supposed to provide precise descriptions of the nature and function of them. Companies must also provide the details of the consignee's business, under C5 of the ELA, as well as the precise purpose to which the equipment will be put, C6. However, in ELA after ELA in the case of machine tools exporters, the consignee's business and the precise purposes of the goods were simply stated as 'general engineering'. Such was the extent of this that the Head of OT2/3 confessed that, 'it seems almost as though we are being parodied or satirised'.[92] The President of the Board of Trade told the House of Commons on 13 June 1995 that a recent DTI survey had shown that 74 per cent of licence applications from arms exporters had not been properly completed.[93]

The DTI stated to the Trade and Industry Committee that 'there is a need for . . . the *bona fide* nature of the end use and user (to be) established',[94] but, at least in the case of machine tools to Iraq, there was a confusion, whether purposefully created or not, between the DTI and other Departments about the meaning of these terms. When the DTI referred to full end use or end user assurances in discussions with the FCO and MOD, what it was referring to was the statements upon the ELAs and not, as the other Departments seem to have thought, separate end user certificates.[95]

While the DTI's mission to promote exports leads it to use its statutory power to act unilaterally where it can to avoid the inter-departmental process of control, the Department nonetheless knows that this is a power it can only use at the margins of the control system. Again, while it might mobilize the companies to appeal refusals and marshall arguments concerning employment and business, neither of these tactics would find success if there were not receptive minds in other Departments. The contest between trade promotion and export control is one that is fought *within* what is formally and informally controlled. In other words, the contest concerns matters of interpretation. If the FCO and MOD were misled about matters of end use and end user, then, again, one will not find any evidence, at least as far as export decision-making for Iran and Iraq is concerned, where the DTI's actions and arguments did not fit the intentions of policy-makers.

The DTI is as sensitive as any other Department to what is and is not the will of the centre. Furthermore, if the exercise of export control is not as rigorous as it could be, this is the responsibility of the Cabinet:

first, for placing control in a Department charged with promoting exports; and, second, for not providing the appropriate level of resources for the staff concerned to pay the proper attention to each ELA. However, for the most sensitive defence exports, the FCO and MOD are involved and can exercise countervailing pressure.

## The Ministry of Defence

### Organization

Whereas in the DTI there is a conflict between roles, but no conflict within the Department, in the MOD, there is an inbuilt tension between both the missions of the Department and the staff who perform them. As Pearson argued:

> Within the MOD, two sets of perspectives seem evident: on the one hand, UK armed services' concerns about the acquisition of equipment and about the release of sensitive technologies to the wrong people abroad – with possible resultant use against the United Kingdom; and on the other, the marketing concerns of the Defence Sales Organisation and those promoting exports of British equipment.[96]

As such, before one can assess what the dominant perspective is in the MOD, it is necessary to establish the separate roles and views of the Department's sub-organizations. What conflict does exist is essentially between the DESO, on the one hand, and the security branches and the DIS on the other. The DESS is supposed to act as 'honest broker',[97] reconciling the views of the two wings of the Department: it is, within the MOD, the 'regulatory or monitoring body'.[98] Export control, however, really rests with the security and operational staff.

### Security Advisers

ELAs from the DTI are sent by DESS to the appropriate security adviser (land, sea, air or industrial). The adviser can go anywhere within the MOD or outside – for example to the Research and Development Establishments or the appropriate Defence Attaché – for assistance. He can also contact the company, either directly or through the DESO or the DTI, for further information. If Lt Col. Glazebrook is an archetypal security adviser, then one can say that the MOD's security staff are scrupulous in the performance of their export control functions, whether they are operating regulations concerned with national security, or guidelines which spring from wider foreign policy concerns.

In the 1980s a sense of frustration was apparent among the security advisers. First, they seemed to be outnumbered by what they saw as the combined forces of the DESO and the DESS; and, second, their function was undermined by there not being a proper infrastructure for control. Lt Col. Glazebrook complained:

> One of the faults of the whole of the MOD's clearance system is that we do not have instant access to a computerised set-up . . . that can tell us how much of a particular category (of equipment) has been authorised over the past two or three years.[99]

This meant that the quota system, used as a tool in preventing diversion, whereby goods are authorized in quantities sufficient to meet the recipient's Armed Forces' requirements but no more, was rendered less effective. It also meant that, at least in the case of Iraq, the UK contributed to the establishment of an indigenous defence industry, by the provision of manufacturing and dual-use equipment, because the *accumulations* of goods were not discerned. Whether or not these frustrations contributed to an unusual level of resignations among the technical advisers in the late 1980s must be a matter of speculation.[100]

### The Defence Exports Services Organisation

The DESO is the successor organization to the Defence Sales Organisation (DSO). The latter was set up in 1966 after a report, commissioned by the Secretary of State for Defence, on the promotion of Britain's defence exports. Amalgamating the various established sales divisions within the MOD, it was not so much the internal reorganization and the creation of a new post of Head of Defence Sales, but the appointment of a businessman to that post which generated, as Stanley and Pearton called it, 'the impetus behind the sales drive'.[101] In the establishment of the DSO, as Pierre said, the UK 'sought to bring the marketing and promotion techniques of private industry to government'.[102] While the organization was renamed in 1980 to reflect 'the secondary, supporting role of government and the primary selling role of industry',[103] its Head remains a seconded appointment from industry and its commercial ethos is unchanged.

The DESO reports to the Procurement Executive and to the Permanent Under-Secretary and the Secretary of State for Defence. Beneath the Head of Defence Export Services are two Under-Secretaries and a two-star officer. The Director-General of Marketing, whose 'job (is) to get out and sell things'[104] heads four regional marketing directorates, in turn divided into specific country desks, who keep in close contact with both industry and Defence Attachés.

A fifth Directorate, Marketing Services, collects export sales data and market intelligence, as well as promoting the export perspective in the formulation of policy on equipment procurement. It is the Marketing Directorate which is responsible for shepherding ELAs through the inter-departmental decision-making process. A Military Deputy heads a team of Armed Forces personnel who give advice to companies, guided by the Regional Marketing Desks; and liaise with the military authorities within the MOD 'to give perspectives on the needs of outside industry'. This is performed particularly in relation to new projects 'even if that means altering specifications'.[105] This team also arranges demonstrations of equipment. A second Under-Secretary and his team, supported by the DESS, 'police policy decisions'[106] and provide administration and support. The DESO also has Offset and International Finance Advisers. The US, Malaysia, Kuwait, Saudi Arabia and Iran have dedicated project offices.

By 1989, the DESO employed about 700 civilian and service staff. Most of the staff 'do not have direct marketing or business experience, although a number of key posts are occupied by secondees from industry'.[107] In an NAO Survey, most companies expressed a wish to see more 'high quality staff from the private sector as well as less frequent staff changes'.[108] Thus the DESO might be seen as something of a hybrid creature: part military, part private industry and part public bureaucracy. Two matters should be noted: the DESO's identification with the interests of the defence companies; and the fact that the secondment of industrial personnel places the defence manufacturer inside government.

A DESS official said of the DESO staff: 'They would get enthusiastic about the product . . . you would not be any good at the job if you did not do that.'[109] Certain implications naturally flow from the DESO's mission. As Sir Richard Scott argued:

> Every advocate knows that you start identifying yourself with your cause, so that the danger is that the government officials will identify with the causes which they are backing as advocates . . . Is there danger down the line for the decisions that will be taken?[110]

In 1975, a member of the Expenditure Committee of the House of Commons said: 'There have been recently suggestions made in this place and in the press that the arms sales organisation has some kind of a policy of its own.'[111] But this 'policy' is its mission. More significant is the fact that the DESO ensures that private industry, and its concerns, are situated inside the administration in that some of its key personnel are seconded from defence companies. The case of David Hastie, a Divisional Marketing Director of the Military Aircraft Division of British Aerospace (BAe) and the project manager of the Hawk fighter aircraft is illustrative.

Hastie was seconded into the DESO for eighteen months in 1988 and 1989. His official title was Business Development Adviser (BDA) to the Head of Defence Export Services (HDES) and his tasks included various marketing projects, including one for the European Fighter Aircraft.[112] However, it was BAe that approached the DESO about Hastie's secondment, although, in fact, the Civil Service Commissioners were told that it was the MOD personnel who contacted the company with a request for a candidate. Furthermore, Hastie, throughout his secondment, continued as a salaried member of BAe's staff with his expenses paid by the company; and the post of BDA was a new one which did not survive Hastie's departure. The Head of the DESS recalled that the proposed sale of Hawk to Iraq, over which BAe had long been negotiating, 'was a consideration' in the appointment. The same official also recalled that Hastie went to a desk in the DESO which dealt with Hawk, and that while he, and not Hastie, 'wrote the main policy papers in that area', he 'took advice from' Hastie.[113]

Hastie was well-informed about Government thinking *vis à vis* exports to Iraq. Indeed, having met Hastie in Baghdad, the British Ambassador telexed the FCO to report:

Certainly much of what Hastie said about recent discussions in London was new to me and my Defence Attaché. It may be that you are already engaged in some review of defence sales policy, but I think we should be brought more into the formulation.[114]

Another event in Hastie's DESO career is also significant. The FCO vetoed DESO representation at the Baghdad Fair in April 1989 even though the MinDP was content for staff to attend. Their fear was that attendance by the DESO 'would risk leading the Iraqis to misread this as a deliberate sign of a change in Her Majesty's Government's policy on arms sales'; and, moreover, the Iraqis 'might also take advantage of the presence of MOD officials by suggesting an official meeting which we would certainly have to refuse.'[115]

Hastie, however, went to the Baghdad Fair, released from his secondment in the DESO in order to lead the BAe team, but returning to his Government post immediately on his return from Iraq. When the FCO learned of his departure, too late to intervene, they were 'alarmed' since his status was regarded as 'ambiguous and open to misinterpretation'.[116] The Head of the DESS recalled that the initiative to send Hastie to the Fair came, not from the company, but from the HDES. Moreover, the MinDP approved the visit.[117]

Hastie's visit to the Baghdad Fair was not his only trip to Iraq on BAe business during his secondment in the DESO. The ambiguity about his position was expressed in a signal, in July 1989, from the Embassy in Baghdad: 'D. Hastie of BAe: is he now Aerospace or is he MOD?'[118] Hastie had been in Iraq to present a revised offer on Hawk. The Embassy reported: 'the technical details were now agreed in principle. The Iraqis had hoped he had brought with him a price for the new offer'.[119] As if to emphasise the problematical nature of Hastie's status, someone in the MOD had underlined this quotation and added two exclamation marks.

The Head of the DESS saw no difference between what Hastie did in Iraq and what the HDES might have done. Hastie, he said, was working as 'an arm of the HDES'.[120] Thus, his secondment – and the manner in which it was done – suggests that this was a possible mechanism whereby the DESO were able unofficially to negotiate with Iraq about the Hawk when such negotiations would have been politically difficult. At the time there was both public concern about Iraq's use of chemical weapons and its abuses of human rights, and the existence of official guidelines which forbade the export of lethal equipment.

Such appointments were justified by the Head of the DESS in the following terms:

> He brings with him benefits as well as gaining benefits. If, in fact, there was any way in which the employment of such a person in government would place his company at an advantage over other UK companies, that would be very much against the rules . . . He should not, I think, be put in a position where he is making policy judgements.[121]

The argument could be made that Hastie, for example, with his place in the bureaucracy and his control of information (his discussions with the Iraqis) had, at best, *indirect* influence over policy judgements. Certainly, within Government, and particularly from the

MOD, momentum built behind the sale of Hawk. This pressure was only stopped by a Cabinet decision. Locating companies inside the bureaucracy would seem to increase the *internal* pressure to make sales. While civil servants within the DESO share the companies' interests, they are also socialized in the British bureaucratic norms of co-operation and compromise, and are aware of the political sensitivities of the Party in Government. Industrial secondments have but one loyalty.

Even though the organization is, in the words of a former Ambassador, 'dead keen to sell',[122] this impetus is meaningless without power. The DESO's power rests upon two pillars: first, it exercises the sorts of arguments used by the DTI in discussion with others; and, second, it exploits the MOD's standard operating procedures to try to get ELAs and AWP applications approved. These pillars, in turn, rest upon the foundation stone of the DESO's influence, which is access. As a former Head of Defence Sales put it:

> The only people who can sell are the people who build and design the equipment . . . So our role really is to sort out the departments within Government – Trade and Industry, Treasury, FCO, ECGD, and so on . . . One important factor is that we do have access.[123]

By definition, it is the Minister who is most receptive to the employment, industrial and trade arguments which the DESO deploys in seeking the granting of export licence and AWP applications. The HDES enjoys frequent access to the MinDP, the Minister primarily responsible for arms exports and control. Lord Trefgarne, as MinDP, reported that he saw his HDES 'twice a week, because I was always wanting to keep abreast of all the things that Sir Colin (Chandler) was active in'. Furthermore, Lord Trefgarne said that he 'would generally say, "Yes"' if the DESO and the company concerned wanted to see him about an export licence.[124]

Like the DTI, the DESO's power of persuasion increased in direct proportion to the value of the ELA under consideration. Another route around uncertain approval was a failure to provide sufficient detail upon the ELA. This was a source of permanent complaint by the security staff who felt that 'things were not being properly looked at because of the lack of information'.[125] If the security adviser did request more information before making a decision, the application would be pended. This had the merit of keeping the customer 'on the hook',[126] but its main advantage was that it had:

a damming effect . . . we would suddenly find that (we) had a dozen or more applications all waiting. At that stage what tended to happen . . . (was that) . . . the case would be taken to Ministers on the grounds that there was a very great deal of money for the UK held up awaiting information.[127]

The DESO's other main tactic was to use the appeals procedure. Like the DTI in respect of ELAs, the DESO is suspected of prompting companies to appeal AWP applications. It is the DESO which communicates the AWP decision to the manufacturer. Lt Col. Glazebrook was concerned that in being 'twice distilled . . . (the communicated decision) missed some important point I was trying to make'. When asked if the AWP could receive copies of the replies to companies, the DESS replied in the negative, citing a lack of resources for such a procedure.[128] The use of the appeal seems to have been a frequently used tactic by the DESO. As Lt Col. Glazebrook put it: 'The DESO have a tendency, if they do not get the answer they like from me, to raise it to my superiors in the hope that they will be kinder.'[129]

In theory, appeals against refusals are to be referred by the DESS to the RMIPC, which is the parent committee of the SXWP. Unlike the SXWP, it is an internal MOD group, although it usually has an FCO representative, and consists of 'five just men, all very high-ranking' who 'reached good and neutral decisions' but who are 'not always in favour of sales'.[130] However, their seniority ensures that weight is given to their view by the Minister. In the case of appeals against refusals for exports to Iran and Iraq, this procedure was not followed, but, according to Lt Col. Glazebrook, 'It has always been difficult to persuade the DESS to follow their own procedure.'[131] Instead, the DESO attempted either to get a different ruling from a superior officer, or a further meeting between the security adviser and one of the DESO's military officers who could dispute the case.[132]

Appeals are supposed to be conducted only if and when the company could provide further information that would support a different decision. That did not happen, at least in respect of goods for Iraq. Instead, in a series of *ad hoc* meetings, chaired by a member of the DESS and attended by the appropriate Regional Marketing Desk who represented the company, the security adviser found himself 'reiterating the same points over again . . . because I have no power on my own unilaterally to say "stop it". An appeal procedure is on and they can go on appealing if they like.'[133]

Such disagreements between the security advisers and the marketing staff were, according to the Head of the DESS, 'most usual',[134] and appeals 'quite frequent'.[135] Finally, the DESO's third informal route of appeal lay in a direct approach, if the DESS agreed, to the Minister.[136]

## The Defence Exports Services Secretariat

The evidence suggests that, in the MOD, the security advisers, who judge potential exports against the framework of policy, are somewhat outnumbered and out-manoeuvred by the sales promotion staff of the DESO. It is the DESS which operates between these countervailing, but unevenly matched, forces. The DESS 'helps to formulate policy', acting as a focal point in the MOD for all the control regimes to which the UK belongs. At the same time it 'helps to ensure that the policy formulated is observed' in that it co-ordinates and conveys the MOD assessment of AWP applications and ELAs. Nonetheless, the DESS remains 'a DESO secretariat'.[137]

A former Head of Defence Secretariat 13, (the DESS's predecessor organization), defended this structural position thus:

> By being part of the DESO, DESS can, in fact, keep in touch with what is going on with the regional marketing directors. It can warn people off when it looks as though some project is being worked up which really does not stand much of a chance of being permitted . . . if the regulatory or monitoring body was outside the organisation, business would be slowed up immensely and . . . would become much more formal, remote and confrontational.[138]

The deputy to the Head of the DESS, DESS 2, said that 'the security advisers regarded us (as) on DESO's side, but, because we acted as a brake on the activities of the DESO, RM desk officers regarded us (as) on the advisory side'.[139]

Lt Col. Glazebrook argued that, within the DESS, 'the more senior you got, the more neutral the line'.[140] In respect of defence sales to Iran and Iraq, it was DESS 2, Alan Barrett, who arbitrated between the security branches and the Regional Marketing Desks. Unlike in the FCO, where submissions to Ministers progress up a recognized official chain of command, Barrett could, if he wished, go directly to the Minister, although he generally discussed matters informally with the Head of the DESS.[141] In the case of machine tools for Iraq, Barrett misrepresented the views of the security advisers to other Ministries. He was, at least in this case, on the side of the DESO. Indeed, it was

Barrett who provided the *DESO's* input into a re-consideration of export policy towards Iraq in 1990.[142]

Like all secretariats, the DESS's power lies in its control of information. Information can be withheld both from those who make the final decision and those who make the initial judgement. Again, in the case of defence exports to Iran and Iraq, the final inter-departmental decisions were withheld from the relevant Working Group in the MOD. The Head of DESS explained this decision, like the decision not to provide security advisers with the DESO's communications to companies, as stemming from the secretariat's excessive workload.[143]

## Minister for Defence Procurement

In any contest between sales and restraints, which the DESS is unable to resolve, the attitude of the Minister is fundamental to the balance of power. Moreover, his perspective indicates to all concerned how he would wish the dispute settled and thus arguments can be resolved without direct reference to him. It is the MinDP who has primary responsibility for both sales and controls. The role of the other Minister of State, the Minister for Armed Forces (MinAF), in the decision-making process seems to have been, at least until 1986, unclear.

In 1986, the MinAF wrote to the MinDP to say that, while the Minster relied upon the MinDP's staff for advice on which sales would be 'materially adverse', he nonetheless sought the power of approval, either by himself or his Under-Secretary of State, 'before clearance of the sale by the Armed Forces' side of the house is given'.[144] The memorandum was written following the MinDP's overruling the security advisers' refusal to countenance an export which they argued would have 'given away a serious vulnerability to British forces'.[145] After this particular intervention, the balance between sales and restraints tilted slightly in favour of the latter. Henceforth, the security branches could appeal to the MinAF, or threaten to do so, if they were overruled on a refusal based upon security grounds. That said, it was still the MinDP who carried the main responsibility for arms export policy.

Whereas Lord Trefgarne, the MinDP between 1986 and 1989, was inclined not to reject recommendations but to ask 'to be convinced',[146] his successor, Alan Clark, was 'a difficult Minister', who 'would not necessarily accept any MOD advice'.[147] Clark was also less accessible to his officials than Trefgarne.[148] Although Trefgarne does not seem to have enjoyed a reputation as a restraining influence upon defence sales, Clark marked a reinforcement of the disposition to sell. Staff were warned, some months after Clark's arrival, that:

the new Minister was approving more defence overseas sales and will not necessarily accept security advice that a particular sale should not take place . . . he was gung-ho for defence sales.[149]

Clark defined his duties as MinDP as being 'responsible for defence exports and for the defence export sales organisation'. As such, his 'attitude was identical' to the one he had had as Minister for Trade, although he thought he 'should probably have been more sensitive, because (he) was better informed as to the security implications'.[150] In contrast, Trefgarne defined his role more broadly to include 'overseeing and supporting' the constraints of Government policy as well as the promotion of defence exports.[151] In what Clark called 'borderline' cases, he 'understood that (his) responsibility in the DESO context should take precedence', relying upon the MinAF to overrule him if 'clear-out security considerations were paramount'.[152]

## Perspectives

Given the nature and overall mission of the MOD, such 'clear-out security considerations' can be expected to be respected, although there is ambiguity in this notion. However, the DESO is more powerful than the security branches in the MOD. Sir Adam Butler, then MinDP, recalled that by 1984 'the defence sales tail was beginning to wag the dog a bit more strongly than it had . . . very considerable support was given to that sales effort at a Ministerial level'.[153]

While, as this chapter has argued, all Departments are politically attuned to the centre, from the MOD's perspective, defence sales serve certain organizational goals. Sir Adam Butler argued that it was important to note that, from the MOD's point of view, there was a 'general policy of selling defence equipment'.[154] The former Head of Defence Secretariat 13 concurred: 'there are a large number of constituencies within the Ministry of Defence, and I think most of them are really quite keen to see a healthy defence export industry thrive'.[155]

That there is this support for arms sales is not simply a function of the DESO's power within the MOD. It is because certain benefits are seen to accrue to the Department as a whole. As a member of the DESS put it: 'the overall feeling was that in the interests of reducing the defence budget, and helping the British economy, the idea was to sell as much as possible'.[156]

Sir Adam Butler summarized the thinking of the MOD: 'if we could increase the sales of equipment which our own Armed Forces were using, it would clearly have an effect on the unit cost of that equipment'.[157] However, Sir Adam later admitted that:

> One of the fundamental industrial arguments is that volume reduces unit costs. Ergo, if you can achieve that in the defence field, then you will reduce unit costs. The reason it didn't apply (to the UK) to any great extent was that we never produced the volume.

Savings in procurement costs only occur with repeat orders and, moreover, the most expensive equipment, where R&D costs have been greatest, is the equipment which Britain has 'generally failed to sell in sufficient volume'.[158] Nonetheless, there is no reason, based as it is upon a self-evident micro-economic principle, why the quest for procurement savings should not remain a goal of the MOD, or, indeed, government as a whole. Furthermore, the MOD, like any other Department, must compete for a share of limited funds from the centre. To the extent that the arms exports effort gives every appearance of an organization attempting to be both efficient and self-reliant, that sales effort is a worthwhile enterprise.

As for constraints, Pearson has argued that 'The influence exercised in favour of arms transfers by . . . private business interests as well as by a government agency charged with promoting sales, contributes to the erosion.'[159] But this is not only to see the power of the DESO, and, by extension, the power of industry, in isolation from the rest of the decision-making process, but also to mistake Government policy for implementational weakness. On the first point, Sir Adam Butler compared the balance between sales and constraints within the MOD 'to a standard corporate decision-making process where one element, one part of the firm's interests, must be restricted according to the overall policy of that firm'.[160]

On the second point, the following internal MOD discussion about constraints upon sales is illuminating. According to a minute written by the Assistant Under-Secretary Commitments (AUSC) in 1990: 'The MOD view is that, in principle, the only relevant criteria for defence sales in general are commercial interest and national security.' The AUSC argued that international obligations had also to be undertaken with 'due regard' for these two considerations. Passing reference was made to 'presentational factors' in relation to 'particularly odious regimes'.[161]

However, in his reply to the AUSC, the Head of Security Overseas Commitments thought it 'difficult to regard all British foreign policy as, in fact, falling within those two motivations' and argued that, even if the MOD accepted these two criteria alone in judging potential sales, 'the nature and public and international perceptions of the regimes and particular weapons would need to be taken into account'.[162] However, he also accepted that such considerations were 'of course for the FCO to argue, rather than the MOD'.[163]

What is evident, even in a Department which is overwhelmingly disposed to sell arms, is the political sensibilities of officials. It is not the case that there is an erosion of control: officials interpret the controls by reference to political interests. The contest within the MOD is not about policy but about testing regulations and guidelines against the conflicting, and often ambiguous, interests of security, commerce and, to a lesser extent, the perception of Britain's international reputation.

## The Foreign and Commonwealth Office

### Organization

Within the FCO, export licence applications come first to the Non-Proliferation and Defence Department (NPDD), formerly the Defence Department. The NPDD ensures that they are passed to the relevant country desk, and, if the ELA is chemical or biological in nature, also to the ACDD. Export licence applications with relevance to nuclear weapons are now considered by the NPDD. The FCO's role is to examine the third of all ELAs which are referred to it in terms of international agreements, national guidelines and other foreign policy considerations. It is also the FCO's role to act as the channel of communication with other states in terms of the co-ordination of policy, and, for example, in issuing démarches regarding particular goods to particular countries. There is 'no systematic clearing house for information among friendly governments' outside of the Australia Group and MTCR frameworks.[164]

Pearson argued that it is difficult to bring foreign policy to bear upon defence export control when 'the Foreign Office lacks technical expertise on weapons systems and must rely on MOD technology assessments or open source material'.[165] However, in the case of ELAs for Iran and Iraq, there is evidence of the FCO challenging MOD assessments. Moreover, it must be assumed that technical expertise resides in the ACDD and NPDD. Both Departments can be directly involved in the decision-making process or can be consulted by the country desks.[166] But, in any case, the role of the FCO is primarily to provide political, rather than technical, input.

Internal disagreements are less profound in the FCO than in the MOD. Although the Department might be described as 'a room of many desks'[167] and, therefore, somewhat fragmented, and although there might occasionally be disagreement between the country desk, concerned to preserve good relations, and either the ACDD or NPDD, both of which are concerned with global issues of proliferation, neither the country desk nor the two Departments 'can ignore the other's firm disagreement'. Disputes are resolved by the Minister of State.[168]

## Perspectives

The main foreign policy concern in assessing potential defence exports seems to be a commercial one, with national security being much more the preserve of the MOD's security branches. There is, however, also a continuing concern for reputation, which is part of the mental architecture of great power status, and can serve to constrain sales. But, it should also be noted that lingering pretensions to such status also serve to promote arms sales as a projection of British power in another form.

Contrast the two views below, the first from Alan Clark when Minister for Trade and the second from a senior diplomat:

I am blighted by the Foreign Office at present. Earlier today a creepy official who is 'in charge' (heaven help us) of South America came over to brief me ahead of my visit to Chile. All crap about Human Rights. Not one word about the UK interest; how we saw the balance, prospects, pitfalls, opportunities in the Hemisphere.[169]

And:

The British interest was usually set in a broader context, particularly the need to maintain the full range of civil exports to the region.[170]

While Alan Clark might be impatient with the Foreign Office – he had hoped that Cecil Parkinson, had he made it to the Department, would have 'shifted it from "diplomacy" to trade promotion'[171] – there is enough evidence to say that commerce is now its primary concern. It is unclear to what extent this results from Conservative Party ideology or a longer standing reorientation of British foreign policy. The Foreign Office knows that 'Britain exports or dies'[172] and this acts as both a constraint upon and a springboard for defence sales. Where the potential recipient is willing to link arms sales to access to the general market of the country, the FCO's concern for commerce inclines it to support sales. The FCO is thus disposed to seek restraint where commercial relationships with neighbours and/or adversaries of the potential recipient will be damaged. This is the Department's most important role in the export licensing process. It advises on the political, and thus the commercial, reaction of other states to particular sales.

In any case, the FCO is as receptive as any of the other Departments to arguments concerning the employment and industrial ramifications of exports. As the Foreign Secretary, Douglas Hurd, said: 'Industry . . . expect the British Government to be energetic on their behalf . . . and the Foreign Office spends more and more of its time doing precisely that.'[173]

## The Export Credits Guarantee Department and the Treasury

The ECGD has 'a statutory duty to support UK exports', although it is supposed to operate its trading activities at no net cost to the Exchequer.[174] This condition has proved impossible to meet with the level of defaulting experienced in the 1980s. When considering the extension of export credit, the ECGD analyzes: the country against a rating system; the buyer, by seeking a report from an agency in the country concerned; and, if necessary, the company requesting support. Credit limits are set for individual countries by the inter-departmental Export Guarantees Committee, chaired by the Treasury, which also considers the provision for large individual defence sales. Where the limit for military exports to a specific country has already been reached, the Committee can decide to provide additional cover, although it must take into account the impact of such a move on the civil export business.[175]

The ECGD does not require sight of an export licence before granting credit, although the policy is invalidated if the company does not subsequently obtain one.[176] Export licence and AWP applications are sometimes sent first to the Export Guarantees Committee 'to test whether ECGD cover would be available'.[177] This procedure again underlines the importance of credit to securing a defence sale.

The ECGD's apparent caution towards arms exports is echoed in the Treasury, whose criteria for assessing military sales include, not only the ECGD's exposure, but also an assessment of:

whether, particularly in the poor developing countries, the alloca-
tion of resources to arms rather than, at least in economic terms,
more productive investment is the right way for a country to be
going.[178]

The Treasury reasons that if a state spends disproportionately on arms,
its ability both to pay its debts and to purchase in the future are imped-
ed, and thus British interests are adversely affected. In this sense, the
Treasury's perception of how the national interest is best served is rather
different from that of other Departments. Its view is a longer one.

## The intelligence agencies

Although the intelligence agencies participate in the inter-departmental REU, by and large, their involvement in arms export
decision-making is *ad hoc*. The Joint Intelligence Committee (JIC) produces a weekly survey of assessed intelligence which is circulated to
Departments and the DIS work closely with the security branches in
the MOD. Otherwise, the intelligence agencies' involvement in the
decision-making process lies in the production of raw intelligence
reports, although the DIS has some capacity for assessing intelligence.
The work of the intelligence agencies becomes pertinent to the defence
exports decision-making process in respect of any diversion of arms
from one country to another, the end user, and the end use of goods
with both civil and military applications.

Although the MOD can request sight of an end user undertaking, its
value is open to question.[179] Alan Clark went further: he would 'never
attach any significance to assurances from . . . practically anyone'.[180]
Of end *use* undertakings, the Director of the Export Licensing Unit
said, in a memorandum to Customs and Excise:

> We have to accept what is said as the only evidence we have of the
> end user. In state run economies there is a recognition that the end
> user could be switched but since we do not exercise territorial con-
> trol, we cannot prevent that happening.[181]

Intelligence, in revealing the customer's real intentions, ought to be
an invaluable tool in negotiating a path between two conflicting interests – the loss of export revenue on the one hand, and diversion and the
diffusion of weapons, particularly those of mass destruction, on the
other. Reports produced by the DIS, Government Communications

Headquarters (GCHQ), the Security Service and the SIS supplement the existing information Departments have at their disposal when evaluating ELAs. This data include ascertaining: whether the quantities and sophistication of equipment to be supplied are 'in line with the stated end use', even if there is no system for keeping watch upon cumulative quantities; whether special inducements are being offered to the company; and the reputation of the country concerned.[182]

However, there are real and abiding obstacles in the way of the effective use, as opposed to production, of intelligence information. The Scott Report highlighted retrieval and dissemination. However, source protection, the attitudes of decision-makers to intelligence and, most of all, the political and economic context within which intelligence agencies must work are more significant barriers. These problems are most conveniently discussed in the later examination of the machine tools cases.

## Centralization and shared perspectives

Neither the bureaucratic politics nor the organizational process model help explain British decision-making about arms exports. On the one hand, the policy process is centralized and, on the other, shared perspectives are predominant over divergent ones. The competition between Departments over arms sales decisions needs to be seen in its proper perspective. That is, it is necessary to examine the process of co-operation. For example, while the DTI exercises its statutory authority to act unilaterally on occasion, in general, it observes the rules of inter-departmental decision-making and respects the veto of the MOD and the FCO even though it might contest their views: the DTI accepts the interruption of exports for reasons of foreign policy. When the DTI suggested that, in view of Iraq's partial trade embargo against the UK, that arms export controls be unofficially relaxed, the Assistant Under-Secretary of State for the Middle East in the FCO took the proposer aside to warn him that, if the DTI 'lost FCO support, they would be very exposed in Whitehall'.[183]

Similarly, the MOD and FCO accept the limits of their power in defence export decision-making.[184] Hence, as a senior MOD official said:

We had a very close relationship, which had to be close and friendly. This whole process was long and fairly cumbersome and we had to maintain the goodwill of our colleagues in other Departments.[185]

The need for inter-departmental co-operation was recognized by the FCO when a senior official noted, in 1988, that:

> There is evidently some way to go in tightening up consultation within Whitehall, particularly with the Treasury and ECGD, over possible economic constraints on major arms sales.[186]

Every Department knows what its role is in the process, and each has a role in both the implementation and formulation of policy. For example, when reconsidering defence exports policy towards Iraq in 1990, each Department contributed, via the Cabinet Office, to the paper prepared for the Cabinet meeting. Moreover, 'if any particular department objected sufficiently strongly to a joint paper, they can put in a document on their own behalf'.[187]

Disputes about the implementation of policy, as well as policy itself, are decided at the centre. The mechanisms for this centralization, which also ensure that implementation coheres with political intentions, are the well-established procedures of communication between Departments and the centre. For example, letters concerning arms sales proposals sent from Minister to Minister are copied to the Prime Minister;[188] and any changes of policy are directed to the relevant Cabinet committee or to the Prime Minister for decision.[189]

The centre gives implementation its direction, not just by the policy it makes, but by the way it both resolves disputes and distributes power. For example, the Treasury defines Britain's economic interest somewhat differently from the DESO, but, as the following extract from an internal memorandum, written by the Head of Treasury Division Defence Management, makes clear, the Treasury knows the wishes of the centre and adjusts its responses to inter-departmental conflicts accordingly:

> You will see from the papers . . . that the Defence Export Services Organisation in the MOD are gung ho to support sales of military equipment to Iraq and almost anywhere else . . . Our main problem is that neither (the ECGD's exposure nor the UK's economic and financial interests) tends to win in Ministerial debate, especially with the Prime Minister.[190]

If the DESO is powerful in arms export decision-making it is because, as Brzoska and Ohlson pointed out, in the 1980s, the 'growing promotional apparatus' was 'headed in practice by the Prime

Minister'.[191] The Prime Minister's interest in defence sales is perhaps reflected in the fact that a quarterly report on worldwide sales progress, country by country, is completed and sent to the Prime Minister by the Director of Marketing.[192] In short, Departments are only influential inasmuch as the centre allows them to be. Furthermore, such influence as any Department has is not fixed: it is dependent upon the policy positions of the Cabinet and Prime Minister.

In respect of defence sales to Iran and Iraq, the guidelines were politically inspired; they were aimed at preserving relationships, and therefore trade, with both countries. It was therefore the FCO which wielded most power with the Prime Minister in any inter-departmental dispute. The DTI complained of the FCO's confidence in overriding them on arms sales to Iran and Iraq by bringing in the Prime Minister.[193] Thus, what Sorenson said of the American political process would also appear true in the British context:

> Politics pervades ... without seeming to prevail. It is ... an ever-present influence – counterbalancing the unrealistic, checking the unreasonable, sometimes preventing the desirable, but always testing what is acceptable.[194]

That this is the case owes as much to the civil service's sensitivity to the centre's political ambitions as it does to the centralization of power. Thus, it is wrong to argue that, 'Although elected officials do enter the decision-making process, their role is more to lend credence and weight to the pursuit of deals.'[195]

A second reason why neither the bureaucratic politics nor organizational process models is useful in locating the essence of decision in British arms exports is that shared, as opposed to divergent, perspectives are more prevalent among Departments. Hence, 'The (Ministerial) meetings where there is a real dispute tend to be small in number.'[196] While it is broadly correct to say that both the DTI and the MOD are keen to promote arms sales for what might be called organizational motives – there is a 'shared interest'[197] – the FCO has sympathy with their perspective, although it is difficult to attribute it to bureaucratic ambitions. Generally speaking, all Departments want to increase the UK's trade. As a member of the Middle East Department put it, 'the job of Britain is to export things wherever it can'.[198] Or, as a senior diplomat observed, 'this is a country that depends on trade, therefore the fewer the restrictions on trade the better, in general, in terms of the national interest'.[199]

Arms sales can work for or against that interest. In the first instance, they make a contribution to trade simply by being sold. But they also nurture relationships with élites in the Third World where most of Britain's arms exports are to be found. Conversely, arms sales can negatively affect trade when they are sold to countries which have neighbours or adversaries which will politically object and commercially express that objection by discriminating against British products. The foremost role of the FCO is to alert the DTI when that set of circumstances might apply. As a former Minister for Trade put it: 'I think the DTI's real interest was, of course, civil goods, and the long term trading relations between lots of countries to whom we export a great deal.' And thus: 'Clearly, one of the factors that (was) in the DTI officials' minds ... was the long-term effect on civil sales of any decisions taken in the defence field.'[200]

There tends, therefore, to be much more of a coincidence of interest than a divergence between the DTI and the FCO. Similarly, the observations of, *inter alia*, Pearson and Brzoska and Ohlson that the FCO 'has been overruled or by-passed'[201] might owe more to a series of incidents involving conflict between the DSO and the FCO over arms sales to the Shah than to a deep-seated and chronic antagonism. A Foreign Office minute from 1988 makes clear the shared interest which results in co-operation between it and the DESO:

We agreed that, at working level, co-operation between DESO and FCO in the past seven or eight years has become very much closer ... The understandable instinct on the part of HDES and his staff to go it alone, which had characterised the 1970s and led to considerable problems, had now been curbed as far as the need to take due consideration of FCO political interests was concerned ... the main thrust of FCO interest is, in my view, to conclude a sale where the political factors are positive rather than to constrain it.[202]

Disputes between Departments occur at the margins, not least because most strategic controls are imposed from without, from the various regimes to which Britain belongs. Conflicts occur where guidelines allow flexibility to be exercised according to circumstances. The only exception is the Treasury whose perspective differs only from the other Departments in that it has a wider perspective on Britain's economic interests.

The argument has been put that:

> ... the government's stated reason for not having rigid guidelines is that it can react quickly to changes in the international environment. Yet this goal is contradicted by a decision-making process in which low-level bureaucrats apply standard operating procedures and work on the basis of precedent.[203]

But this is not the case. Low-level bureaucrats do not apply standard operating procedures. The contest between Departments is about the testing of regulations and guidelines against what is politically acceptable according to the myriad circumstances in which the thousands of export licensing decisions are taken. Departments manoeuvre *within* the framework of policy, not without.

## Decision-making and motives for arms exporting

Finally, having established both that policy is dictated by the centre and not by the bargaining of competing organizations, and that implementation accords with political intentions, the final task here is to analyze what the decision-making process reveals about the UK's motives in exporting arms. Pearson argued that, 'There is a severe divergence between the control of foreign policy and control of defence sales in the British government.'[204] This is reflected in the fact that:

> The Foreign Office plays a distinctly secondary and reactive role in decisions to promote or proceed with sales . . . it is not the main clearing house for obtaining inter-Ministerial approval or comments on sales.[205]

It is accurate to say that the Foreign Office's role is reactive but, short of making the FCO assume the DTI's role in export licensing, or, conversely, the DESO's role in arms export promotion, this will necessarily be the case. The FCO does not play a secondary role in decision-making. Indeed, it is, where guidelines on defence sales are politically inspired, 'the determining Department . . . the adjudicator'.[206]

From this evidence, Krause was wrong in concluding that 'Difficult decisions are referred to a Cabinet Committee, which again suggests that the political implications of sales enter the process after commercial decisions.'[207]

If analysts fail to see motives of foreign policy in British arms exports it is because they have the narrowest conception of the former. Foreign policy is driven by the national requirement for physical and economic security. Foreign policy, as the pursuit of physical security, enters the UK's thinking about arms sales in the various control regimes to which it is a party. As the pursuit of economic security, foreign policy both promotes and constrains arms sales. The hand of the FCO might not be evident in the promotion of sales, (although this overlooks the role of embassies), because it is *another* Ministry to which that task falls for reasons of expertise, but it is evident in the constraint of exports.

Analysts have rarely explored the linkages between defence and civil trade, for which reason they have missed the role of foreign policy in British arms export decision-making. As Sir Ivan Lawrence of the Foreign Affairs Committee of the House of Commons remarked: 'It has always been the Foreign Office view, that we have interests of an economic kind which outweigh practically every other interest.'[208]

Good political and commercial relationships are so intertwined that it is almost impossible to separate them, and arms sales are part of such relationships. To illustrate the point, Treasury Ministers were overruled by other Departments and, more significantly, the Prime Minister, when they opposed increasing the ECGD's exposure to the Omani market by extending credit for the purchase of patrol boats in early 1990. The Treasury official concerned reported the economic arguments against the provision of credit were overcome by *both* economic arguments of a different nature and political reasons.[209] Certainly, Defence Attachés considered 'political factors' as 'the most significant influence in . . . the UK's willingness to sell arms'.[210]

A mixture of motives thus seems apparent in arms sales decision-making, the most significant of which seems to be trade. Interestingly, the foundation of the Defence Sales Organisation seems to owe more to this broader objective than any precise concerns about the health of the defence manufacturing base. Pierre commented that in setting up the DSO in 1966: 'Britain's balance of payments was then in especially bad straits, and this was uppermost in the minds of Harold Wilson and Denis Healey, then Minister of Defence.'[211]

This is not to say, however, that the well-being of defence firms is not often a consideration in individual export licensing decisions. Certainly, all Departments seem susceptible to such arguments. In this context, the bureaucratic politics model has something to say about British arms export decision-making. As Steel commented:

Participants in the national security bureaucracy . . . feel obligated to serve the national security interests of their country. But the vagueness of what that interest requires or allows impels them to substitute more concrete criteria. The domestic political consequences of foreign policy actions is one such standard.[212]

However, such 'substitutions' take place at the periphery, where judgements between interests are finely balanced.

Thus, Britain's arms export policy is altogether more complex than analysts have previously perceived. This is because arms exports decisions are made, not just in the context of the defence manufacturing base, but in the context of Britain's other national interests. There is a mixture of both economic and political motives which can be difficult to separate. The decision-making process is thus designed to allow those, often competing, interests to be tested against one another.

# 3

# The Guidelines on defence sales to Iran and Iraq: policy and implementation

## Official and unofficial policy

In trying to define the British Government's motives in selling defence equipment to Iran and Iraq in the 1980s, one needs to explain, on the one hand, an official policy of restriction of arms supply and, on the other, the fact that both countries received arms from UK companies. The latter cannot be regarded simply as a failure of control mechanisms. This assertion is based upon a body of evidence which includes: the testimony and accompanying documentation of ex-participants in the arms trade with Iran and Iraq; the motives of the UK in *restricting* arms exports; and the process by which decisions were made.

Although there was a system for reviewing export licence applications for both Iran and Iraq, and officials were, as Sir Richard Scott said, 'conscientious',[1] arms reached both countries because the centre, in defining that system's scope, excluded from officials' remit the scrutiny of ELAs for other destinations which might divert British conventional arms to either country.[2] Furthermore, in excluding potential diversion, the Government knew that both Iran and Iraq were using conduits for arms supply. In a confidential meeting between Sir Richard Scott and the then Prosecuting Counsel in the Ordnance Technologies case,[3] the former remarked that 'The powers that be knew that Jordan was being used (as a conduit for arms to Iraq) to some extent, yes. More than suspicion. They knew it was happening.'[4] The documentation underpinning his view is to be found, he said, in the Arms Working Party minutes and papers of the Defence and Overseas Committee of the Cabinet.[5]

In short, the Government had an official policy of restraint of arms sales to Iran and Iraq and an unofficial one of supply. The two policies require separate examination, both for the sake of simplicity and because the motives surrounding the official policy, where the greater evidence exists, provide a foundation for the discussion of the UK's covert supplies of arms to Iran and Iraq.

A number of possible motives for the totality of British arms export policy towards the protagonists in the Gulf War suggest themselves. These include the general economic and political objectives discussed in the earlier chapters as well as more specific goals. These specific goals include a mix of long-term strategic objectives, for example, securing Gulf stability, either by preventing victory for one side or by prolonging the war; and short-term political goals, for example, influence aimed at securing the release of British hostages. The contention here is that Britain's illicit arms sales to Iran and Iraq complemented the official policy of restriction. The restraint of arms exports to the two countries was aimed at preserving relationships and trade with the Gulf as a whole.

When Iraq invaded Iran in September 1980, the British Government, having declared its neutrality, announced that it would not sell lethal equipment to either participant. In 1984 this policy was apparently strengthened by the introduction of the following Guidelines:

1. We should maintain our consistent refusal to supply any lethal equipment to either side.
2. Subject to that overriding consideration, we should attempt to fulfil existing contracts and obligations.
3. We should not, in future, approve orders for any defence equipment which, in our view, would significantly enhance the capability of either side to prolong or exacerbate the conflict.
4. In line with this policy, we should continue to scrutinise rigorously all applications for export licences for the supply of defence equipment to Iran and Iraq.[6]

Official machinery was instituted to execute the policy. The documentation surrounding the Guidelines reveals that the UK sought, via this formal position, to preserve and enhance its influence in the Gulf, but the overwhelming preoccupation was with the protection of civil trade.

The Guidelines on defence exports to Iran and Iraq should not be thought of either as a palliative for domestic disquiet about the war, and the UK's role as a supplier of arms, or as a morally inspired set of constraints. The Iran–Iraq War, the longest conventional war of the twentieth century, claimed a million lives. For the most part, it was attritional, with Iraq on the defensive from 1982. Chemical weapons were used against Iranian human wave assaults, often involving boys as young as thirteen. Civilians were attacked in wars of the cities. Gulf shipping became a target of both sides. But the Guidelines were meant only to symbolize the UK's position, *vis à vis* the war, to all states in the Gulf.

As such, what was important was the declaratory, rather than the operational, policy. There was therefore no failure in control regarding those British arms which reached Iran and Iraq via third countries, since the Guidelines were a device for identifying those exports which, if *officially sanctioned*, would antagonize either side in the conflict.

## The origins of the Guidelines

The nature of the UK's arms defence export policy towards Iran and Iraq – that 'the motivation (was) the presentation'[7] – is apparent in the origins of the Guidelines. According to the legal notion, enshrined in the 1907 Hague Conventions on Neutral Rights and Duties, a declaration of neutrality requires states to refrain from supplying arms to belligerents. The UK was not alone in interpreting this international legal requirement somewhat differently.[8] For the UK, neutrality equated to even-handedness. Douglas Hurd, then Minister of State at the FCO, argued that while there was a body of international law regarding neutrality which had to be respected, it was not 'terribly clear'. The principle, he argued, 'does not prevent you totally from supply but your neutrality has to extend to the way in which you treat requests from the two belligerents'.[9]

The transfer of all 'lethal' defence equipment to both protagonists was officially banned by the British Government. 'Lethal items' were interpreted as 'military equipment designed to kill',[10] although no formal definition of 'lethality' was ever drawn up by the Government to assist its officials in making decisions about export licence applications for Iran and Iraq.[11] While guns and ammunition were not supplied, directly or officially, some necessarily subjective judgements were made about all other forms of military equipment. For example, two ships, supplied by the UK to Iran in early 1984, were re-fitted with beds and operating rooms and called 'hospital' ships, although they were originally ordered as amphibious assault vessels and had mounts for four 40mm guns. A second example concerns a contract negotiated by International Military Services (IMS), (until 1991 a Government-owned company which sought and managed arms deals), to build an integrated weapons complex for Iraq. This was agreed with 'the most swingeing Ministerial guarantee'[12] on 27 July 1981.

Communication systems and radar were exported and weapons training was regarded as 'non-lethal'. However, the most significant 'non-lethal' export was that of spare parts, especially for tanks, aircraft and ships. This was particularly important as far as Iran was concerned since, of the two belligerents, it had been the greater recipient of British arms. Also of importance to Iran was that the Government had waived the necessity for export licences for repair work. Only when the Guidelines were introduced was the waiver revoked.

One should not therefore look to international legal obligations as an explanation for the UK's behaviour, but rather to the balance of interests and the uncertain state of relationships with the two countries. When the Gulf War began, Iran was already the subject of sanctions imposed by the European Community in response to the detention of American diplomats. Sanctions came into force with the Export of Goods (Control) Order 1980 of 29 May and were lifted by the UK on 22 January 1981. While relations with post-revolutionary Iran were uneasy, the UK's relationship with Iraq was also a difficult one at the beginning of the Gulf War. This was in most part due to Iraq's close formal relationship with the Soviet Union and its Ba'ath ideology – anti-monarchist, socialist and pan-Arabist – so inimical to other Arab states with which Britain has close ties.

In attempting to disrupt Iran's war effort, the US Administration began 'Operation Staunch' on 14 December 1983. The State Department instructed its embassies, in countries believed to be transferring arms to Iran, to encourage their host states to desist from the trade.[13] The US, its relationship with Iran shattered by the revolution and the existing hostage crisis, had become concerned by Iranian counter-offensives which began in early 1982. By the spring of 1983, the Administration believed that the Gulf War had turned in Iran's favour.

In February 1984, the Pentagon publicly criticized British transfers of spare parts for tanks and aircraft to Iran, warning that even small items could be 'incredibly helpful'.[14] In July of the same year, the US again criticized the UK for continuing to supply Iran with 'non-lethal' defence equipment;[15] and, when the Prime Minister visited Washington in early 1985, she was urged to withhold arms transfers to Iran.[16]

The then Head of Defence Secretariat 13 (DS13) remembered that the Guidelines were in part a response to American concerns about the refurbishment of British-supplied hovercraft and the supply of Olympus gas turbine engines for British-built frigates for Iran: 'The Americans had thrown in their lot with Iraq. The Arabs had always been very pro-Iraq. We were coming under intense pressure really to abandon Iran altogether.'[17]

The Desk Officer for Iran in the FCO recalled 'a major series of exchanges with the Americans'.[18] In a note to Adam Butler, MinDP, Richard Luce, Minister of State at the Foreign Office, referred to pressure from both the Americans and Iraq's friends in the Arab world.[19]

The Government, in reviewing its arms export policy to Iran and Iraq, had three choices: 'abandon Iran altogether'; institute a total embargo; or keep or modify the existing policy. The first was little discussed. The MinDP wrote of Iran at the time: 'We would not wish to totally sever the residual defence links which we have with her . . . it is important that relations of a more general sort are maintained with her.'[20] While the Americans and Arab states were concerned that Iran was making significant gains in the Gulf War, their alarm does not seem to have been shared by the UK.[21]

Honouring the existing contracts with Iran was seen as necessary in order 'to be able to restore good relations both politically and from the commercial point of view as well'.[22] There are, however, more references to the *commercial*, rather than the political, importance of Iran to the UK. The Foreign Office Minister, initially a supporter of a complete embargo, emphasised the contractual obligations to Iran under the Shah and the consequences of not fulfilling them: 'we would be liable to compensate British industry to many millions of pounds'.[23] By January 1984, Britain had agreed with Iran that thirty-six of seventy-four outstanding IMS defence contracts would be completed. Seventeen contracts referred to spare parts for Chieftain and Scorpion tanks.[24] Hence, the second guideline – the fulfilment of existing contracts – would allow trade with Iran to continue in the face of US and Arab pressure.

There was some discussion of instituting a total embargo, on the initiative of the Foreign and Commonwealth Office, but this option was dismissed. While it would resolve the issue of American and Arab pressure to cease trading with Iran, it would antagonize Iraq and its supporters. Iraq, however, wanted the UK to do more to support its war effort:

> . . . the Iraqi regime, with whom Britain at the time had considerable non-military – ordinary commercial – relations were very irritated with us in Britain for not following other nations, Arab and non-

Arab nations, in selling arms to them.[25]

More importantly, Iraq's Arab supporters influenced the decision not to institute a total embargo:

> we had important defence obligations towards them . . . in the general sense of contracts and training commitments, and, therefore, they took a very keen interest in our policy and particularly the possibility of there becoming any sort of arms embargo.[26]

The Foreign Office Minister accepted in the course of 1984 that a ban on all sales was not possible because of the 'likely adverse reaction from Arab states to the ending of all defence sales to Iraq'.[27] Thus the third guideline – the only new part of the UK's policy – was introduced to appease American and Arab concerns about Iraq's apparently faltering war effort. The Foreign Office Minister wrote to the Head of the Middle East Department in December 1984:

> Although there are new Guidelines largely concerned with the presentation of policy, they also represent a modification of the substance of that policy . . . They are more acceptable to Arab countries and the US Government. They will give rise to greater difficulties with the Iranians and to British defence equipment manufacturers on the other hand.[28]

As the Head of the Middle East Department commented, 'There was a change, but not a major change.'[29]

## Impact

It is difficult to assess the effect of the Guidelines upon Iran and Iraq in military terms. Only a few general observations are possible because of the lack of data. Although the Government released a list of items exported to Iraq between 1987 and 1990,[30] no such list has been issued for Iran. Only a partial picture of approvals and refusals has emerged. While some military training was allowed under the Guidelines, precisely what was permitted is not known. Iran was allowed only 'non-military' Service training; Iraq was permitted training so long as it did not involve lethal weapons instruction or that which was 'politically sensitive'.[31] Iraqis attended courses at the Royal Military Academy, Sandhurst. Pilots and aero-nautical engineers from Iraq were also trained in the UK.[32]

|  | Received | Approved | Refused | Cancelled/ Withdrawn |
|---|---|---|---|---|
| **1985** | | | | |
| Iran | 259 | 218 | 10 | 31 |
| Iraq | 367 | 323 | 11 | 33 |
| **1986** | | | | |
| Iran | 298 | 256 | 17 | 25 |
| Iraq | 295 | 272 | 6 | 17 |
| **1987** | | | | |
| Iran | 593 | 405 | 61 | 127 |
| Iraq | 315 | 272 | 16 | 27 |
| **1988** | | | | |
| Iran | 607 | 457 | 51 | 99 |
| Iraq | 401 | 327 | 25 | 49 |
| **1989** | | | | |
| Iran | 568 | 396 | 71 | 101 |
| Iraq | 410 | 345 | 19 | 46 |

Source: Letter from the Department of Trade and Industry to the Trade and Industry Committee, *Exports to Iraq*, Minutes of Evidence, Session 1991–92, House of Commons Papers, 86–viii, p. 297.

Table 2: Number of applications and outcomes for Iran and Iraq, 1985–89

As Table 2 shows, over a five-year period, 1985–89, 8 per cent of the total applications for Iran were refused and 4 per cent for Iraq. Iran accounts for a total of 2,325 ELAs and Iraq 1,788. ELAs for Iran thus exceed those for Iraq by 30 per cent. The total of approved ELAs for Iran was 1,732; 193 more than Iraq at 1,539. Thus, 12.5 per cent more applications were approved for Iran than Iraq in this period. Two notes of caution are necessary here.

First, the low rate of refusals for both sides can be accounted for by companies seeking clearances in advance of ELAs. There are no equivalent figures for AWP clearances for Iran and Iraq. Second, there is only partial data about the value of defence exports to Iran and Iraq in the 1980s. Between 1980 and 1984, Iraq imported £184 million and Iran just £13 million of British defence goods.[33] British defence exports to Iraq were worth £35 million in 1985 and £33.2 million between May 1986 and April 1987.[34] In 1988, defence sales to Iran were worth just £1 million.[35] The values for 1989 and 1990 are given in Table 3.

| | Approved | Refused |
|---|---|---|
| 1989 | | |
| Iran | 50 million | 70 million |
| Iraq | 130 million | 20 million |
| | | |
| 1990* | | |
| Iran | 120 million | 5 million |
| Iraq | 75 million | 1 million |

\* Figures are for the first half of 1990 only.

Table 3: ELAs approved and refused for Iran and Iraq by value[36]

However, these figures should be treated with some caution in terms of judging Britain's policy: the figures represent demand as well as supply. In the early 1980s, relations with Iran were particularly tense: sanctions were applied to Iran in 1980 and pre-revolutionary contracts were the subject of conflict. The very low figure for Iran in 1988 might reflect the UK's policy *vis à vis* the war, or it might reflect Iran's 'Buy British Last' campaign, instituted in 1987. In looking at the comparative figures for 1989, it should be borne in mind that Iran was being treated much more severely than Iraq because of the *fatwa* on Salman Rushdie. The picture changed quite markedly in the first half of 1990.

The partial data about the values of defence exports to Iran and Iraq give the impression that there was a tilt to Iraq. Moreover, there were 'asymmetrical results'[37] in the application of the Guidelines; for example, the supply of radio equipment did not constitute a 'significant enhancement' to Iraq, because it had good communications, but did so to Iran because it did not.[38] Thus Guideline (3) favoured supply to Iraq over Iran. However, it should be noted that the UK supplied to Iran equipment which was of importance to the war effort. The 'tilt' to Iraq in Guideline (3) was mitigated by the 'tilt' to Iran in Guideline (2): the continuing supply of spares, under existing contracts especially benefited Iran, given the export, before 1980, of major weapons systems from Britain. Before the outbreak of war, Iran had received from the UK: some 875 Chieftain main battle tanks, 250 Scorpion light tanks, a destroyer, four frigates, six landing ships tank, a replenishment ship, military hovercraft, Seacat naval missiles, 114mm naval guns, and Rapier and Tigercat surface-to-air missiles. The major British weapons systems Iraq had received before the war consisted of fifteen Hawker Hunter and two Heron aircraft, forty Gazelle, seven Puma and seven Lynx helicopters and Milan anti-tank guided weapons.[39]

## Implementation

It was only when the Prime Minister endorsed them, on 12 December 1984, that the Guidelines came into force.[40] They were first announced to Parliament on 29 October 1985 by the Foreign Secretary. Special institutional arrangements were put in place to scrutinize export licence applications for Iran and Iraq. The chief forum for this scrutiny was the Inter-departmental Committee on Defence Sales to Iran and Iraq (IDC). Represented on the IDC were officials from the DTI, the MOD, and the FCO, which chaired and produced the minutes of the meetings held more or less every month. Each Ministry had an informal veto on the approval of applications and an informal right to refer refusals up to Ministers. Summary records of meetings were submitted to Ministers for approval, with particularly sensitive items highlighted.

It was the MOD which had to assess ELAs and AWP applications on their 'lethality' and whether or not they 'would significantly enhance the capability of either side to prolong or exacerbate the conflict'. This task fell to the MOD Working Group (MODWG) comprising: the Regional Marketing Desks for Iran and Iraq, respectively RMD2A and RMD2B, from the DESO; and operational, security and intelligence staff from within the MOD. The MODWG was chaired by a representative of the DESS, who took the committee's recommendations to the IDC where they could be overruled. The final outcomes of the IDC were not reported to the MODWG.

The MODWG's task was essentially a technical one. To illustrate the point, a member of the security staff represented on the Group, Lt Col. Glazebrook, described how he assessed the Matrix Churchill machine tools. One particular batch was capable of producing 500,000 155mm shells. With information supplied from Defence

Intelligence regarding the number of 155mm guns in each division of the Iraqi Army and the number of divisions engaged in combat, as well as the expenditure rates, Lt Col. Glazebrook calculated that the enhancement was insignificant. He initially characterized the machine tools as 'lethal'.[41] The MODWG continued to assess ELAS for Iran and Iraq against the traditional restraints on sales – military security, operations, proliferation and the transfer of advanced technology – as well as against the Guidelines.

Had the Guidelines truly been formulated as a response to the international obligations of neutrality, or as a mechanism for ensuring an early end to the war, one of the UK's declared policy objectives,[42] then one would expect the process of evaluation to end with the MODWG. Sir Richard Scott is worth quoting at length on this point.

> It is not easy to see why it was necessary to have an IDC at all except that . . . the IDC provided another committee where the views which would otherwise have been out of court, because the MOD operational security would have ruled them out of court, are given a second bite and can then argue, with whatever hope of success they have . . . for the licences to go.[43]

At the IDC, MODWG assessments were challenged by either the DTI or the FCO. This reflected the deliberate subjectivity of the Guidelines. In essence, the IDC allowed ELAs to be tested against the balance of British interests. While the Chair of the MODWG represented the MOD view, if allowed to speak, the DESO representatives, sometimes outnumbering the rest of the members of the IDC,[44] argued the case for British military exports. The DTI asserted the interests of British trade and British companies, while the FCO advised the other Departments upon 'the political factors implicit in the new criteria'.[45] That is, the FCO was able to say what exports to Iraq the Iranians would tolerate and what sales to Iran the US and the Gulf would accept. The MODWG provided the basis for justification of decisions to the various interested parties. The IDC's other purpose was to 'speed up decision-making',[46] a reflection, perhaps, of the importance Iran and Iraq placed upon defence purchases and the UK's desire to be seen to be helpful to both sides.

Officials understood the Guidelines in different ways. The security and operational staff of the MODWG toiled faithfully to apply the Guidelines, conducting careful military evaluations. They clearly believed that the Government did intend the belligerents to be denied lethal items and equipment which would enhance their capabilities. In the FCO, a comment made by one official in 1984 indicates that he also believed the Guidelines to be a sincere attempt by the Government to take a just course in the Gulf War. This is one of very few references to morality in the documents thus far made public:

> The Americans and Arabs may wish we were tilting more to Iraq, but we have consistently and publicly emphasised our evenhandedness and impartiality in this conflict. I think it would be both inadvisable and morally wrong to be tempted down the primrose path of short-term expediency.[47]

The official's superior had a keener sense of the purpose of the Guidelines. He replied: 'I can understand the principles of expediency and "balance" in arms sales, but I get a bit confused when morality is involved.'[48]

When the Guidelines were set aside, when MODWG assessments were challenged, it was not because officials were pursuing a different agenda from that of their political masters: it was because they understood the nature and purpose of the policy. Before examining that in greater detail, it is important to note other facets of the policy process.

Policy was formulated and later re-formulated by the centre. The Guidelines, drafted by junior Ministers, became the framework for policy when the Prime Minister approved them. A Cabinet Committee, MISC118, oversaw policy towards Iran and Iraq. The instruction to officials to ensure that civil trade with Iran was not affected by the operation of the Guidelines came from the Foreign Secretary,[49] and was reiterated, from time to time, by Ministers to their officials.[50]

Sir Richard Scott found that the changes in the Guidelines, following the ceasefire in August 1988, 'were not put to senior Ministers for approval',[51] and that the first the Prime Minister knew about the revisions was in July 1989.[52] The paper trail certainly leads to this conclusion: but a consideration of the circumstantial evidence tends to a quite different explanation.

On 31 August 1988, the Foreign Secretary sent a paper to the Prime Minister, and members of the Overseas and Defence Committee of the Cabinet (OD), entitled, 'The Economic Consequences of an End to the Iran–Iraq Conflict'. The paper asserted that:

> Our defence sales policy will need to be reviewed . . . we can use discretion within the Ministerial Guidelines to adopt a phased

approach ... relaxing control on a growing number of categories as peace takes hold.[53]

The Prime Minister approved the strategy and asked to be 'kept very closely in touch at every stage and consulted on all relevant decisions'.[54]

FCO officials then circulated a draft paper which added detail to the desired 'phased approach' by suggesting that dual-use goods then held up be released.[55] On receipt of the paper, the Foreign Secretary decided that he did not want it circulated 'and thereby initiate a process whereby it will become known that our line on defence sales to Iraq has relaxed' because 'it could look very cynical if so soon after expressing outrage over the Iraqi treatment of the Kurds we were to adopt a more flexible approach to arms sales'.[56] Other Departments came to know of the Foreign Secretary's concerns at the IDC meeting on 19 October 1988.

On 28 October the Foreign Secretary, still wishing not to circulate the FCO paper, nonetheless authorized officials 'to operate flexibly within the Guidelines'.[57] In the meantime, given the Prime Minister's wish to be kept informed, junior Ministers were copying their correspondence about the proposed revision of the Guidelines to her. One such letter, dated 14 November 1988, from the Foreign Office Minister, William Waldegrave, to the Minister for Trade, Alan Clark, noted their agreement to 'adopt a more flexible attitude' towards the Guidelines.[58]

On 21 December, the three junior Ministers of the FCO, DTI and MOD met to revise the Guidelines. The new wording of Guideline (3) was copied to the Foreign Secretary and the Secretary of State for Trade and Industry, with a letter to say that, after consultation with Gulf Posts and Washington, the Ministers would recommend that the Foreign Secretary put the revision formally to the appropriate Secretaries of State and the Prime Minister.[59] The process was overtaken by the *fatwa* on Salman Rushdie. The IDC reverted to the stricter Guidelines for Iran pending the Cabinet's decision about sanctions:[60] the new Guidelines had been used ostensibly on a trial basis for both countries from the beginning of February.[61] Junior Ministers confirmed the decision to treat Iran under the old Guidelines on 24 April 1989.[62] The Foreign Secretary was informed on 28 April.[63] He had made clear to the Prime Minister on 7 March that senior Ministers did not favour a defence exports embargo against Iran.[64]

Sir Richard Scott argued that the Prime Minister was not informed of these changes. He noted that Ministers ceased to copy their corre-spondence to her after the Foreign Office Minister's letter of 14 December 1988.[65] That letter communicated the decision by junior Ministers to meet on 21 December to revise the Guidelines.

Given that everyone in the decision-making process knew of the Foreign Secretary's emphasis on relaxing the Guidelines without it becoming public, and had, with the Prime Minister's knowledge, authorized a *de facto* relaxation in his Note of 28 October, it is not impossible that the failure to put the revision to senior Ministers was an agreed device for fulfilling the Foreign Secretary's wants. The policy with regard to Iran, post-*fatwa*, had also to be agreed informally by junior Ministers to similarly avoid initiating a process whereby the change would become known.

With regard to the Prime Minister's knowledge, it is hard to think of an explanation for the sudden and simultaneous end to the copying of the three junior Ministers' correspondence to her. The most plausible interpretation is that, having been informed of the meeting to revise the Guidelines, to send succeeding papers to her would mean bestow-ing official status on the changes and, again, initiating the process the Foreign Secretary feared. Witnesses insisted that the Prime Minister would have been kept informed of changes in the Guidelines.[66] More-over, many have testified to the informal, as well as formal, ways of communication within Whitehall. Lord Howe, the former Foreign Secretary, said that no minutes would be kept if he was simply giving information to Ministerial colleagues, including the Prime Minister whom he saw once a week. Whitehall is, he said, 'a mutually sensitive culture'.[67] By way of further circumstantial evidence, civil servants have been told, post-Scott, that any warnings given to Ministers that their actions might be unlawful must never be put in writing.[68]

Another instance of the mutually sensitive culture at work can be found when the DTI pressed for a revision of the Guidelines in 1990. A lengthy review process, co-ordinated by the Cabinet Office, and involving all Departments, was embarked upon. This process culmi-nated in an inter-departmental meeting, chaired by the Foreign Secretary, who, in turn, sought the Prime Minister's approval for the recommended revision that only lethal equipment be denied Iran and Iraq.[69]

It should also be remembered that Ministers' approval of IDC rec-ommendations was required. Sensitive equipment was highlighted or brought to their attention; and where a Department exercised its veto, Ministers were required to take decisions. Four export licence applications for the two countries had to be resolved by the Cabinet.[70]

The faithfulness of the organizational process to the policy of the centre is demonstrated by the following incident. At a meeting in December 1985 with Tariq Aziz, Iraq's Deputy Prime Minister, the Prime Minister had made an inaccurate 'off the cuff remark'[71] that the UK 'had terminated the supply of weapons to Iran and even the supply of items which might possibly have a military application'.[72] The FCO drew the remark to the attention of the IDC. Given its nature, on the one hand, and the existence of the Guidelines 'as the prime source of policy' on the other, officials interpreted the Prime Minister's comments as meaning that they should 'err on the side of strictness' in assessing future ELAs.[73]

An understanding of the UK's motives behind this policy can only be attained by penetrating the nature of the Guidelines. Disputes, particularly between the MODWG military experts and the defence sales staff, arose around Guidelines (2), 'existing contracts', and (3), 'significant enhancement'. What constituted 'lethal equipment' (Guideline (1)) continued to be debated as it had since the beginning of the Iran–Iraq War.

The DESO persuaded others to extend the definition of an 'existing contract' to cover contracts signed *after* 1 January 1985. Having argued 'backwards and forwards', the Chairman of the MODWG asserted, with support from the DESO Regional Marketing Desks, that: 'Where we have granted a licence and (the company) had subsequently signed a contract . . . we should continue to meet the contract even if circumstances changed.' Guideline (2) did mean that military spares continued to be supplied to both countries as companies would normally be contracted to supply them over the lifetime of the equipment originally supplied.[74]

The argument that existing contracts had to be honoured was also deployed when it became known that Matrix Churchill machine tools, already issued with an export licence, were being used to make artillery shells in Iraq.[75] Also used, in partial justification for allowing machine tools to be exported to Iraq, was the argument that these goods were dual-use items and unless contrary *evidence* was provided, the DTI had no reason to refuse a licence. The onus for providing evidence was placed firmly upon the MODWG.[76] This argument was assisted by the vague, or deliberately misleading, way in which applications were completed, particularly that part which referred to 'end use'.

The argument regarding evidence was also deployed in respect of other dual-use items. Aircraft spares for dual-purpose aircraft were therefore permitted. Spares for Iranian Boeing 707s, F-27s and C-130s were approved at various times in spite of the use of these aircraft in operations in the Gulf. The DESO insisted that 'of course they (the aircraft) would be used for civil purposes at all stages'[77] and since civilian versions of aircraft used in military operations were also owned by the two countries, the military experts were forced to concede defeat.

The third Guideline was obviously problematical, not least because Iran and Iraq 'were two very different countries fighting a very different type of war'.[78] Although Defence Commitments Rest of the World (ROW) provided the MODWG with assessments of the state of the battle between Iran and Iraq, judgements were necessarily subjective. Furthermore, the MODWG knew that Guideline (3) 'was meant to be flexible'.[79]

Even when the MODWG concluded that a particular item would represent a 'significant enhancement', it could always be second guessed by the IDC, aided – significantly – by the fact that these were subjective judgements and *ipso facto* open to interpretation. For example, an ELA for a Marconi Tropospheric Scatter communications system for Iran was said to be for civilian use. When the MODWG dropped the points where it was to be located on to a map, it was discovered that these points represented the main military centres in Iran. The MODWG was concerned that it would be used to 'control operations in the Gulf against the Armilla Patrol' and that it not only represented a 'significant enhancement' but also a threat to British forces.[80] The application, worth £40 million, was, however, permitted, after an appeal by the company pursued on its behalf by the DESO. This example serves to show the two strategies used by the DESO and companies when seeking to export defence equipment to Iran and Iraq. The first was to declare equipment to be bound for civilian use, where plausible, and then to put the onus for showing otherwise on those who objected. The second was to emphasise, or exaggerate, the value of particular contracts.[81]

However, one should not be misled into thinking that the DESO was in some way subverting British policy. The Guidelines, as noted, existed as a framework for the examination of the balance of British interests. In such cases as the one referred to above, what was being weighed was the trade interest on the one hand, and the political, and therefore opposing trade, interests on the other. The licence was approved by Ministers in 1990, nearly two years into the ceasefire. In other words, there was little threat to the Armilla Patrol at that point so what had to be weighed against trade was the likely reaction of the

Arab states of the Gulf. Since the FCO could argue, with some plausibility, that the British Government had believed the communications system to be for civil purposes, the scales were left tilted towards trade or the company's interest. As Lord Howe, then Foreign Secretary, argued, 'foreign policy considerations were more likely to be invoked . . . if there was any publicity'.[82]

It is this balancing act which captures the essence of decision-making within the Guidelines as it does for all British decision-making about arms exports. Interests are weighed against each other. As Alan Clark said of the Guidelines, 'They were high sounding, combining, it seemed, both moral and practical considerations and yet imprecise enough to allow real policy considerations.' In short, the Guidelines represented 'a brilliant piece of drafting'.[83]

While decision-making is necessarily devolved downwards, junior Ministers had to approve the decisions of officials and, at the same time, were well aware of both the range of their power and which decisions needed to be referred upwards. Furthermore, as the then Foreign Secretary recalled, it was 'the Prime Minister and I who were the fontees or originees . . . of the policy'.[84] The whole process was set up to allow competing interests – those 'real policy considerations' – to be tested in inter-departmental fora because decisions about arms exports were not straightforward or obvious. Certainly, officials and Ministers competed to have their Department's view prevail, but all participants were ultimately aware of the policy preferences of the centre and the interests they share are more noteworthy than those over which they diverge. For example, the DTI accepted the restriction on the defence trade with Iran and Iraq because it perceived the Guidelines to be in its interests in promoting civil trade. As an official from the DTI recalled, the Guidelines 'preserved the UK's neutral position . . . and enabled us to go on exporting very large amounts of civilian goods to both countries'.[85] The DTI accepted that the Guidelines could not be withdrawn, in 1989, because of the potentially hostile reaction of Arab countries and a subsequent loss of trade.[86]

## Companies, public and Parliament

Before leaving the decision-making process for defence exports to Iran and Iraq, the role of actors outside Whitehall should be considered: the companies, Parliament and public. It is very difficult to disentangle a company's interest from that of the country, since the latter clearly has an interest in the well-being of the former. There are numerous examples of individual companies being able to exert pressure upon deci-sion-makers either because of the value of the contract, viz Marconi, or because of the delicate state of the company's trading position, viz Matrix Churchill.

But such examples should not distract one from the fact that, for the most part, defence companies could not export their goods to Iran and Iraq. In spite of the large profits at stake to BAe, in a deal worth between £1 and £3 billion, the proposal to sell fifty Hawk aircraft to Iraq was turned down in July 1989 because of the other interests which would have been jeopardized by its sale, notably relations with Iran, with Arab Gulf states and with Israel,[87] as well as adverse public opinion. The proposal might also have been turned down because the Treasury finally won the argument about the extension of credit to Iraq, upon which provision the deal ultimately rested.[88] Hard and fast rules about the influence of companies are therefore difficult to make. However, the fact that British companies were forced to forgo trade with Iran and Iraq while foreign companies supplied the belligerents might have been a factor in the decision to trade illicitly.

The 1984 Guidelines were *not* a response to public or Parliamentary pressure. Concern about the use of chemical weapons by the Iraqis had been largely done away with when the Government introduced the Export of Goods (Control) (Amendment No. 6) Order which made licensable eight chemicals which could be used for the manufacture of chemical weapons.[89] Sir Adam Butler, MinDP, recalled that in 1984 and 1985:

There were relatively few direct approaches to . . . Ministers, (I probably dealt with most of them) . . . there were no Parliamentary Questions to myself on the subject . . . there were very few letters dealing with it . . . it was not a subject which was being debated in Parliament.[90]

Periodically there were Parliamentary Questions about British poli-cy but no Debates on the issue. In 1984 and 1985, there was no evident public or press outrage at the Government's policy of continuing to supply 'non-lethal' defence equipment to both Iran and Iraq.

The Guidelines were not, then, a sop to domestic public opinion. The pressure to change the UK's policy came from outside, the United States and from Iraq's Arab supporters. It was because of these pres-sures that an announcement to Parliament was delayed for some eleven months.

Occasionally, officials and Ministers were warned to be careful about the 'public defensibility of individual decisions'. One such occasion was after revelations of secret American arms sales to Iran.[91]

Again, in August 1987, the UK sent four British Hunt-class mine countermeasure vessels and a support ship to join the Armilla patrol because of the increased danger to shipping from mines. Consequently, there was increased media attention paid to the Armilla patrol. A FCO memorandum referred to the media coverage and noted that:

> It would not be publicly understood if it were known that we are continuing to sanction the export of spares that were enabling the Iranian Navy to mount attacks on our own and other friendly ships.[92]

Similarly, following the Rushdie affair, the IDC was concerned about the 'major presentational difficulties involved in any large defence exports to Iran';[93] and a licence application for a microwave detector was deferred 'because of the possibilities of public misinterpretation of the uses of (it)'.[94]

Public opinion occasionally brushed against the Government's policy, nudging it towards temporary restraint, but the main thrust of policy remained unchanged. FCO internal guidelines say that 'close attention should be paid to the recipient country if such a country has a bad record of human rights',[95] but there are no references to the human rights records of either Iran or Iraq in considering export licence applications, except in relation to public opinion. However, even after 'tens and tens and tens of letters about the gassing of Kurds and political prisoners',[96] pressure of public opinion did not change the policy: it merely made the Government more circumspect about *some* exports. But, more than anything, it made the Government anxious to prevent disclosure.

Parliament certainly did not scrutinize the Government's policy on defence sales to Iran and Iraq. The Guidelines were subject to changes in interpretation over the period and these were not disclosed to Parliament. These changes could be 'formal', that is 'instructions in writing to do something different from the Guidelines as published in Parliament',[97] or informal and unwritten. However, both variations were treated in the same way. None of the changes, detailed below, were reported to Parliament. Indeed, the timing of the announcement of the Guidelines in 1985 sprang from a tactical need to head off criticism about the export to Iran of armoured vehicle spares.[98]

In 1984, a ban on chemical warfare equipment for both sides was introduced. In mid-1986, Ministers instructed officials to take 'a softer line with Iran'.[99] In September 1987, the Foreign Secretary directed that spares for the Iranian Navy were to be stopped, because of its increased threat to the Armilla patrol. By early 1989, this ban had been lifted. There is also a reference to a change in the Guidelines for Iran because of 'evidence of Iran's involvement in terrorist activities',[100] although what form this change took is not evident. At some stage in the conflict, greater attention was paid to the transfer of technology to Iraq because of 'the increasing degree of collaboration between Iraq and the Soviet Union'.[101] Most controversial of all, in terms of accountability, has been the changing of the Guidelines, preceded by a more relaxed interpretation of them, following the ceasefire in 1988.

At a ministerial meeting, in December 1988, the third Guideline was amended to read:

> We should not in future approve orders for any defence equipment which, in our view, would be of direct and significant assistance to either country in the conduct of offensive operations in breach of the ceasefire.[102]

This change simply reflected the changed circumstances of the ceasefire: it was a relaxation of British policy. However it was not to become official policy because of public sensitivity following Iraq's use of chemical weapons against the Kurds. Following the crisis in Anglo-Iranian relations caused by the *fatwa* on Salman Rushdie, Ministers confirmed the IDC's decision to treat Iran under the stricter Guidelines, but they agreed that 'this need not in principle preclude all sales to the Iranian Navy or to the Iranian Revolutionary Guards Corps'.[103]

Finally, on 19 July 1990, Ministers agreed to relax further the Guidelines although they deferred putting recommendations to the Prime Minister, with her agreement, because of the 'situation between Iraq and Kuwait'.[104] A note of the meeting also refers to the deferral as being 'connected with our bilateral relations with Iran'.[105] Because of a revision of the COCOM controls, those at the meeting did agree to release the relevant dual-use goods.[106]

While the Guidelines had no legal force, their political status remains a matter of debate. The Government subsequently took the view that they represented 'procedure'[107] or a 'tool of policy';[108] and therefore their application, and changes to them, did not have to be announced to Parliament. However, at the time, the Government referred to them as policy. There is truth in both positions.

The Guidelines did represent *declaratory* policy but, at the same time, they were a mechanism for allowing foreign and trade policies to be conducted. They were, in the latter sense, a procedural means to a policy end. It is beyond the scope of this book to judge whether or not Parliament should have been kept informed. The fact that relationships with other countries were bound up with the Guidelines probably means that their interpretation and progress should not have been disclosed. The point is, as the Scott Report made clear, that Ministers failed to consider the public interest in the disclosure of information.[109] It is worth noting that Parliament only pursued the Government after the fact. Parliament was not a factor in the decision-making process, save, like public opinion, as an ephemeral force for restraint where exposure was a possibility.

# 4

# The Guidelines on defence sales to Iran and Iraq: motives

In considering the UK's motives for its defence sales policy towards Iran and Iraq, three broad headings suggest themselves, although they are overlapping and intimately bound up with one another in each serving the physical and economic security of the country. These headings are: regional stability; the preservation and enhancement of relationships with other countries; and trade. Disentangling the last two is no easy task. First, defence and civil exports might be considered as part of the fabric of good international relationships, both a reflection, and a cause, of friendship and alliance. Second, outside the East-West confrontation, it might be true to say that countries seek influence primarily for trade. However, with the UK's imperial past, the quest for influence might well be part of that continuing *psychological* tradition. Unravelling the motives in this particular case is made more difficult by the fact that defence equipment was both supplied and withheld. It should also be borne in mind that such an examination of motives with regard to official policy must, at best, be partial. It is only when the totality of British arms export policy towards Iran and Iraq is considered that one is able to draw firmer conclusions. Nonetheless, the official policy had clearly definable purposes.

## Stability in the Gulf

It is not possible to make clear judgements about the UK's defence sales policy towards Iran and Iraq *vis à vis* a British desire to influence the course of the war because, as noted above, one needs to examine the full range of equipment supplied by the UK to both belligerents. At this stage of the analysis, all one can say is that the official policy did not impact upon notions of stability in terms of ending the conflict, although the Guidelines were presented as addressing that very concern. On this, a former Foreign Office Minister argued:

The Soviet Union, France and increasingly the United States decided that they would give active support to Iraq in this war through supplies of equipment. We took the view, however, contrary to almost every other nation in the world, that we should have an evenhanded, even approach to the war, that it was *not in the interests of the Gulf that this war should be prolonged.*[1] (emphasis added)

The FCO defined the Government's policy objectives in respect of the Iran–Iraq War as:

1. To promote an early negotiated settlement.
2. To ensure that the waterways are safe thus to contribute to upholding freedom of navigation for all merchant shipping.
3. To limit the spread of Soviet influence in the strategically vital Gulf area.[2]

Two opposing lines of argument are plausible. On the one hand, the case could be made that, given the existence of so many arms suppliers, whatever the UK did or did not sell would be marginal in terms of the progress of the war. An FCO official noted in late 1984 that the Guidelines 'will not have any appreciable effect on the military balance whatsoever'.[3] Another FCO official recalled: 'there was scope for disagreement about what difference (an item) made through a war which involved very large numbers of people, huge amounts of equipment'.[4]

On the other hand, had the termination of the Iran–Iraq War been a genuine desire, or, more properly, the *chief* objective, of the British Government then, the line would have been drawn between civil and military equipment, even though, in an age of total warfare, the supply of any product will enhance capability. No distinction would have been necessary between lethal and non-lethal equipment or between goods which 'significantly enhance' capability and those which do not. Had the Government intended to fulfil its objective of not prolonging the war, it would, as an FCO official put it, 'have attempted to draw up a list of equipment, saying: "You may send this, you may not send that"'.[5]

Sir Adam Butler, then MinDP, did not recall correlating the application of the Guidelines to the state of the battle,[6] although that did happen – that was the MODWG's task – but for reasons only indirectly connected with the war. Moreover, when the FCO and the MOD discussed the introduction of the Guidelines, they noted that 'care would be necessary not to exclude sales of all equipment, irrespective of their likely contribution to the prolongation of the war'.[7] The Guidelines, as the head of the Middle East Department said, 'were not intended to be a policy towards the war'.[8] Thus, their only contribution towards regional stability, if we allow that the UK's definition of stability did include the termination of the war, might conceivably lie in the third objective identified by the FCO. That was, in improving its relationships with Iraq and Iran, the UK was seeking to counter Soviet influence.

## International relationships

The UK's arms export policy towards Iran and Iraq was conditioned by the desire to maintain relations with *both* countries as well as with the Arab states and the US. To take the US first, DTI officials complained to each other about 'cases where decisions not to issue licences were taken primarily with regard to US objections'.[9] When the US Administration was revealed as having covertly sold arms to Iran, officials noted that they could not 'expect to exploit' the situation, since the Americans were 'more likely to be rigidly purist from now on, but the events should encourage a slightly lower level of sensitivity (in the FCO) to US feelings'.[10]

An example of the effect of the US on British arms export decision-making for Iran and Iraq concerns Iran's proposed importation of small boats in 1985. These were considered to be a 'significant enhancement'. However, the IDC minutes of 8 January 1985 noted that the Middle East Department 'appeared to be against the supply of any boat or ship, even dinghy, to Iran because of the presentational aspect to America'.[11]

On the same side of the equation were the Arab Gulf states. For example, in 1986, Westland applied for an ELA to refurbish two hovercraft which the Iranians were using to resupply their bridgehead in Iraqi territory. The matter had to be resolved by the Prime Minister, as did the licensing of small boats, after the DTI exercised its informal power of veto. Her response was that: 'We must adhere to the normal guidelines in these matters. Not to do so would mean going back on the assurances which the Prime Minister has given on many occasions to Arab countries.'[12]

A letter from the MinDP to the Foreign Office Minister in 1984 argued that, when briefing British embassies, 'we should be careful not to give too much detail on what is likely to continue to dribble to

Iran'.[13] Thus it was that the Guidelines were not announced to Parliament for some eleven months because:

> If you state publicly that the contracts with Iran were continuing you would be drawing attention to it as a specific in a way which was potentially embarrassing (to) Her Majesty's Government because of our relationship with the Gulf states and with America.[14]

In short, the declaratory policy was aimed at the US and Arab states as a means of making the continuing supply of defence equipment to Iran more palatable. As Michael Heseltine, then Secretary of State for Defence, wrote in a memorandum to the Foreign Secretary in 1984: 'I welcome your assurance that there will be a private degree of flexibility varying and improving the presentation of our policy without radically altering its substance.'[15]

The Guidelines were correlated to the state of the war between Iran and Iraq, but only because it alerted British decision-makers to what defence sales to Iran would and would not be acceptable to the US and the Arab states. The third Guideline was thus a code: deciphered, 'significant enhancement' meant 'Arab and American objections to the sale of equipment to Iran'. In the other direction, the term was equally useful in assessing which officially sanctioned defence sales to Iraq would antagonize the Iranians.

Before considering the various factors which gave the US and the Arab states an influence over British official arms sales policy, it is worth pointing out that the policy reflects the fact that Britain expected even small-scale exports to be noted by other countries. It was perhaps the case that the US Central Intelligence Agency (CIA) which supported 'Operation Staunch', was passing on such information about UK defence sales to Iran to the Arab states of the Gulf as a means of applying additional pressure on the UK.

### The United States

As the UK's most important ally, the reason why the US was able to exert influence over British defence sales might seem rather obvious, especially in the 1980s. As David Sanders said of Conservative thinking:

> Reagan's America was still the 'number one' defender of the capitalist values and interests of the West. Even if Britain no longer possessed the capability to be 'America's *universal* number two', it

could at least provide assistance with all the resources at its command in certain limited contexts. What was good for American foreign policy was good for the capitalist West. And what was good for the capitalist West was good for Britain. This was the essential logic underlying Thatcher's Atlantic circle foreign policy.[16]

Moreover, in the Gulf region there were shared interests between the US and Britain: in containing Soviet influence; in ensuring 'the continued existence of political structures which will ensure an uninterrupted supply of crude oil at stable prices';[17] and in maintaining freedom of navigation in the Gulf.

The US's policy towards the Gulf region was summarized in the Carter Doctrine, which declared that the US would intervene should an 'outside force' attempt to wrest control of the Persian Gulf.[18] However, US policy, in the absence of sufficient American force in the region, relied upon the regimes of Saudi Arabia and Iran in keeping Soviet influence, and other hostile forces, at bay. In 1979, American power, already diffracted in the Gulf, was severely undermined by the loss of Iran to revolution. At this time, the US had no diplomatic relations with Iraq and viewed the regime with some suspicion. As the former US Secretary of State put it, America's 'support for Iraq increased in rough proportion to Iran's military successes'.[19] The US believed, at various points in the conflict, that Iran would win and that, in the words of Donald Rumsfeld, US Envoy to the Middle East in 1984, 'the Gulf could cave in to Iran – a collapse'.[20] In sum, as the House of Commons Foreign Affairs Committee put it, the US saw 'the threat to their "vital interests" as coming mainly from Iran'.[21]

Although America could tilt towards Iraq, extending credit, technology and, later, intelligence, it could not openly supply arms to Iraq, not least because of the US relationship with Israel. Thus, because its policy instruments were so limited in the Gulf, its main tool for influencing the course of the Iran–Iraq War became 'Operation Staunch'. In other words, cutting off Iran's military supplies was viewed as essential to US and Western interests. Thus, on the one hand, the UK's relationship with the US seems to have been a significant factor in restricting the UK's defence exports to Iran. But, on the other, given the closeness of the Anglo-American relationship and the importance the US attached to 'Operation Staunch' in preventing an Iranian triumph, what is perhaps more remarkable is the fact that the UK continued to trade with Iran. This would seem to indicate either a divergence of view about the war or the existence, for the UK, of some countervailing factor.

## The Arab states

While it was the importance of the US as the UK's superpower ally which translated into American influence over British military exports, one must look to other factors in explaining the sway of the Arab Gulf states. Like the US, their fear of Iranian victory led them to support Iraq and, in 1981, to form the Gulf Co-operation Council (GCC), comprising a Defence Committee and a small joint force. However, two points need to be borne in mind. First, the six states – Saudi Arabia, Kuwait, Bahrain, Qatar, the United Arab Emirates and Oman – are not a cohesive group: historical enmities and disputes exist between them; and their attitude to Iran varied considerably. Kuwait and Saudi Arabia had no diplomatic relations with Iran and were openly active in support of Iraq, whereas the lower Gulf states were more conciliatory in their relations with Iran and their support for Iraq was much more covert. Second, support for the regime of Saddam Hussein in the war against Iran was tempered by suspicion of his capabilities and intentions.

The UK has retained strong links with the Gulf states, a region in which it has only recently (1971) relinquished a permanent presence. Treaties of Friendship exist between the UK and Qatar, the United Arab Emirates (UAE) and Bahrain. Defence relations with Oman, Kuwait and Saudi Arabia were described by the FCO as 'close'.[22] It was these formal and informal connections which in part led to the despatch of the Armilla patrol in October 1980, although there are no commitments to provide direct military support in any of the Treaties. Because of the 'extensive personal and trade links' between the UK and the Gulf, according to the Foreign Affairs Committee, 'the British have never "withdrawn" from the Gulf area.'[23] That the UK is eager to retain and increase its influence with the Arab Gulf states is not in doubt. As the FCO put it:

> It is an important British policy objective that we should continue to be regarded as a reliable political interlocutor and trading partner by those states and the government spares no effort in nurturing these relationships.[24]

Certainly, under Mrs Thatcher, Britain's activism in the Gulf increased, although what part circumstances and what part orientation played remains to be seen. She was the first British Prime Minister to visit the states of the lower Gulf, and lamented Edward Heath's failure, when

Prime Minister, to reverse the withdrawal East of Suez. In her memoirs, Lady Thatcher argued that, 'Repeatedly, events have demonstrated that the West cannot pursue a policy of total disengagement in this strategically vital area.'[25] The Armilla patrol indicated that sense of renewed British activism in the Gulf: it also served to increase British influence in the area.[26]

Exports of UK defence equipment were *restricted* to retain influence, not with the recipients, but with third parties. In exporting military goods to Iran and Iraq the Government wanted, as the MinDP said in 1986, 'to ensure that due weight is given to the sensibilities of the Gulf Co-operation Council'.[27] But the UK was engaged in a political balancing act in which the supply of defence equipment, while restricted for reasons of influence, was not ended since the UK sought to retain relationships with *both* sides in the Iran–Iraq War.

## Iraq

British defence sales to Iraq were affected by concerns of foreign policy in two ways: first, equipment was *supplied* to the country in order to gain influence; and, second, goods were *withheld* if Iraq's domestic and international actions displeased, or concerned, the British Government. Although, with the outbreak of war, the risk diminished, the close relationship which existed between Iraq and the Soviet Union meant that export licence applications continued to be assessed in terms of their potential diversion to the Eastern bloc. But Iraq's use of chemical weapons, its human rights abuses and, more particularly, its intimidation of Iraqi dissidents in London in October 1988 and the execution of the journalist, Farzad Bazoft, in March 1990, all impacted upon the UK's supply of defence equipment.

While the Government did use the export licensing system for reasons of foreign policy, the effects were short-lived. However, as in the case of Iran, the withholding of some defence exports has to be seen in the context of restrictions already in place. The case of Iraq's use of its embassy in London for the intimidation of political opponents, which resulted in the expulsion of four diplomats, served only to delay the DTI in seeking a more thorough review of the Guidelines. Furthermore, it was only one factor in that delay.[28] Similarly, the impact of Mr Bazoft's execution was as temporary as it was weak. The only demonstrable effect was the suspension of UK military training provided to the Iraqis.[29] A DTI memorandum noted only that:

In the light of the Bazoft execution and the consequential straining of relations with Iraq, FCO officials are adopting more rigorous scrutiny than hitherto of applications they consider to be sensitive.[30]

Again, a review of the Guidelines, which the Cabinet had wanted, was delayed because of the execution – but only by a few months.[31]

The life sentence imposed upon a British businessman, Ian Richter, in 1986, does not seem to have had any impact upon defence sales. More importantly, no restrictions were imposed as a result of increasing evidence about Iraq's procurement network. To what extent Iraq's use of chemical weapons and its abuses of human rights would have impacted upon the UK's defence export policy *without* the accompanying public outcry remains to be seen. Richard Luce argued, after the International Red Cross confirmed Iraq's use of chemical weapons, that he 'could not see how, under present circumstances, we could justify a policy of tilting in favour of Iraq'.[32] While the Government did more to control chemical precursors and other equipment to Iraq, and Iran, the conventional defence trade continued. The main restriction upon that trade was the UK's desire to maintain a relationship with Iran.

Nonetheless, one should not obscure the fact that the UK did use the export licensing system to express its foreign policy concerns. Ministers decided to relax the Guidelines, in 1988, not only in the light of the ceasefire but also because, after the UK's condemnation of Iraq's behaviour *vis à vis* its use of chemical weapons, 'there had been a willingness from both sides to warm UK Iraqi relations which had produced a slight post-ceasefire rapprochement'.[33] It is clear from the documents that the UK sought influence via its defence trade with Iraq. When considering relaxation of the Guidelines in 1989, officials weighed 'presentation' on the one hand against 'increased British influence' on the other.[34] Explaining the UK's consideration of the proposal to sell Hawks to Iraq, David Gore-Booth, of the FCO's Middle East Department, said, 'We were looking for ways to gain influence over the Iraqis.'[35]

Iraq made clear the importance it attached to securing British defence exports. Richard Luce went to Iraq in 1985 to try to defuse Iraqi criticism of the UK's position on defence sales.[36] It was 'one of Tariq Aziz's missions' to exert pressure upon the UK *vis à vis* arms sales when he saw the Prime Minister in 1985.[37] The Minister of Trade and Industry faced a similar complaint when he visited Iraq after the war.[38]

It is possible that the pursuit of influence with Iraq sprang from a long-standing desire to move Iraq out of the Soviet orbit – British arms were sold to the new *Ba'ath* regime of 1963 with that specific purpose in mind – or from a related effort to secure Iraq's oil reserves. However, the more persuasive evidence points to the motive of trade. Influence was sought in order to improve access to the Iraqi market.

## Iran

Motives of foreign policy are equally evident in what the UK supplied to and withheld from Iran. Anglo-Iranian relations were uneasy throughout the 1980s and beyond. Restrictions on the defence trade were imposed by Britain in response to: Iran's use of terrorism; its activities in the Persian Gulf; and the *fatwa* upon Salman Rushdie. More general sanctions were applied to Iran after US Embassy staff were taken hostage in 1979. However, there seems to have been no retaliation, via the export licensing system, for the beating of a British diplomat in Tehran, even though it led, in the late spring of 1987, to the withdrawal of almost all Embassy staff on both sides.

It is difficult to judge what form restrictions took with regard to Iran's involvement in terrorism or even when such restrictions were introduced. The only documentary reference to any specifics of Iran's use of terrorism was mention of 'the possibility of some Iranian hand in the bombing of the Pan Am flight 103'.[39] The consequence of this possible involvement for British defence export policy was simply that any revision of the Guidelines was not to be announced.

Foreign policy, as national security, seemed to work to restrict sales of equipment to the Iranian Navy. Following the attack on the British ship *Gentle Breeze* in September 1987, the UK closed Iran's Military Procurement Offices in London. However, both acts by the UK were half measures at best. For example, spares for Iranian Boeing 707s, used for refuelling fighter aircraft, were approved in April 1988[40] and the Military Procurement Offices appeared to have continued their work, albeit with a reduced staff.[41]

Furthermore, it remains a moot point whether the 'constant'[42] supply of spares to the Iranian Navy for its British-made frigates and logistic ships before 1987, as well as the sale of, *inter alia*, Plessey mobile, long-range radar, constituted a threat to the Armilla patrol. However, since Iran suspended attacks on shipping when Iraq did so, and attacks by Iran on British shipping were few (seven in all), it might be argued that the Iranian Navy did not represent a threat to British forces in the Gulf. Nonetheless, the example serves to underline the involvement of conflicting interests – operational security versus political influence and economic security – in arms export decisions.

The biggest rift in post-revolutionary Anglo-Iranian relations was caused by the *fatwa* imposed upon Salman Rushdie on 14 February 1989. The week before Ayatollah Khomeini's death sentence upon the author, the Foreign Secretary had met his Iranian counterpart to confirm the full restoration of diplomatic relations. As Lord Howe recalled in his memoirs, following the *fatwa,* 'There was now no option but to put that relationship back into the icebox.'[43] The UK's defence export policy towards Iran was affected by the changed relationship. The Cabinet Committee which dealt with the Iran–Iraq War, MISC118, considered the possibility of imposing a defence sales embargo. The IDC reverted to the stricter Guidelines for Iran that had obtained before the ceasefire[44] and deferred all ELAs pending that decision by Cabinet, (demonstrating again the sensitivity of officials to political circumstances).

By April, the Cabinet had decided that the stricter interpretation of the Guidelines would apply to Iran and that sales to the Iranian Navy and the Revolutionary Guards Corps which posed a direct threat to the Armilla patrol would be refused.[45] The FCO were divided about such a measure: some officials argued that, 'it is logical to keep the diplomatic Rushdie episode separated in itself, and in its consequences, from the export control issue'. Others, however, argued that 'defence sales are always recognised as affected by, if not a reflection of, international diplomacy'. However, it is worth noting both that the effect of so acrimonious a dispute was somewhat limited and that one consideration for the Government was 'the PR aspect'. It would be difficult, argued the MOD, 'to explain why, against the background of a break in diplomatic relations and the vituperation against the UK, we should continue to export defence equipment'.[46]

The reversion to the stricter Guidelines was to be treated with the utmost secrecy: even companies were not to be told in the confidential way that they are normally informed of changes of policy.[47] Since the Government was being forthright in its public statements about the *fatwa*, one can interpret the discreet tightening of export control in two ways. Either the increased restrictions reflected a desire to avoid domestic criticism; or the action was meant to be a non-provocative, but firm, signal to Iran.

The state of the defence trade did follow changes in the relationship. In 1986, after a meeting between officials in Tehran, the FCO argued that 'We should not accept the Saudi veto when Iran is suddenly looking interesting and making overtures.'[48] When relations with Iran began to improve after an Exchange of Notes settling compensation claims regarding damage to each other's embassies, Charles Powell, the Prime Minister's Private Secretary, wrote to the Foreign Secretary to say:

> The general strategy (for restoring diplomatic representation) . . . will obviously require decisions over the next few weeks and months on a number of difficult and sensitive issues, such as the Guidelines for defence sales.[49]

Defence exports were seen as vital in maintaining a relationship with Iran, as was apparent from the risks the UK ran in terms of its relationships with the US and Arab states, in continuing the trade. Iran, like Iraq, made it a condition of a working relationship which was reflected in the overt pressure applied to the UK.[50] Why the UK should want influence with Iran is an important question if one is to be more precise about British motives in supplying defence goods.

First, while some in the US Administration feared that Iran was moving towards the Soviet Union, this does not appear to have been a consideration which the UK entertained. While Robert McFarlane, US National Security Adviser, suggested, in 1985, to the disgust of both the Defence Secretary and the Secretary of State,[51] that Iranian 'moderates' be cultivated by *inter alia* supplying arms,[52] Britain seems to have been more interested in fostering a more general relationship. However, some participants have insisted that the UK supplied defence equipment to Iran because of the hostage crisis in the Lebanon. But, unlike the US, no direct exchange seems to have been sought.

The former Foreign Secretary, the Minister of State at the FCO, a desk officer in the Middle East Department, and Lord Trefgarne who served as both Minister for Trade and Minister for Defence Procurement, all argued, in their evidence before Sir Richard Scott, that the British hostages held by *Hizbollah* between 1984 and 1991 were a factor in defence exports policy towards Iran.[53] The same factor, according to the participants above, also limited what could be sold to Iraq and necessitated extreme secrecy about changes to the Guidelines in case the Iranians perceived such changes as benefiting its enemy.

There are several explicit references to this factor in the documents. Charles Powell, in a Note to the Prime Minister on 1 September 1988, referred to the hostages in the related contexts of Britain's diplomatic relations and defence trade with Iran.[54] An FCO briefing for the Ministerial meeting to revise the Guidelines in 1988 argued, like the above Note, that Iran's support for those holding the hostages was a reason to be cautious about relaxation of the Guidelines.[55] The briefing prepared for the Foreign Secretary, in July 1989, on the sale of Hawk to Iraq argued that 'Approval will inevitably attract criticism from Iran . . . It might also affect Iran's attitude to the British hostages in Lebanon.'[56]

The inter-departmental *Iraq Note* made the following observation in relation to a proposed revision of the Guidelines in 1990:

> The need to treat Iran and Iraq equally over the sale of defence equipment reinforces the need for caution. A public relaxation of the Guidelines for Iraq, but not Iran, would have adverse consequences for our relations with Iran, would make the restoration of diplomatic relations difficult and would threaten prospects of release of hostages.[57]

While the issue of the hostages moved up the domestic political agenda throughout the mid- to late-1980s, especially after the kidnapping of John McCarthy in 1986 and Terry Waite in 1987, the Government insisted that it would not deal with terrorists or kidnappers. However, the statement above, as well as the evidence of Ministers before the Scott Inquiry, constitutes some evidence that the Government did temper its defence export policy towards Iran with the hostages in mind. However, there were so many other factors which necessitated both caution and apparent evenhandedness, that it is difficult to discern its precise effect.

The Government did not retaliate against Iran through the export licensing system for the killing of two British hostages in April 1986. Indeed, in July of that year, the Government was becoming more relaxed towards exports to Iran because of a warming of relations. Similarly the holding, without trial, of Roger Cooper and Nicholas Nicola does not seem to have had a restraining effect upon defence exports to Iran. The UK's inaction in these cases might be evidence of a desire to protect the safety of the remaining hostages in Lebanon as well as the British prisoners.

## Trade

Foreign policy operated upon Britain's defence exports in the form of national security concerns *vis à vis* Iraq's relationship with the Soviet Union and Iran's use of terrorism and its activities in the Persian Gulf. Again, there is evidence of foreign policy concerns in the UK seeking influence with the two countries, as well as with the Arab states, in both withholding and supplying defence goods. However, in terms of the former, it should be borne in mind that difficulties in the relationships between the UK and Iran and Iraq had somewhat short-lived effects in inhibiting the defence trade and those effects were more likely to occur where the risk of exposure to public criticism was greatest. That changes to defence export policy were ephemeral could be the result either of their having achieved the desired end or the fact that they represent gestures rather than meaningful tools of influence.

In supplying defence goods, the UK sought to maintain its relationships with the two countries. However, there remains the question of whether the UK sought influence via the defence trade or was *influenced* to continue the supply of equipment by the recipients. There also remains the question of what, in seeking influence, the UK hoped to achieve. The contention here is that the UK sought influence for the sake of its civil and defence trade and, at the same time, was influenced to continue to supply military goods by the threat, exercised by both Iran and Iraq, to interrupt normal civil trade.

Before examining the evidence which supports that contention, it is necessary to be clear about the classification of these motives. The pursuit of trade, while it straddles the ever-decreasing divide between foreign and domestic policy, ought to be considered as the former: that is, the motive to sell defence equipment might be economic, but the pursuit of trade is nonetheless a primary goal of foreign policy. The defence trade is both an end in itself and a means to an end.

In order to explain fully the UK's official policy towards the Iran–Iraq War, it is necessary to identify its interests in the Gulf. As Pridham argued, 'Britain in the 1980s (saw) the Gulf in terms of oil, finance and commerce.'[58] While the Gulf retains both 60 per cent of proven oil reserves and has the lowest cost of production, the UK is itself a net exporter of oil. In the 1980s, Britain did not rely upon the region's energy resources. However the UK's interest in Gulf oil remained because of 'oil sharing' schemes operated by both the European Community and the International Energy Agency. The UK would have to forgo some of its oil production to support other countries if a disruption to supplies resulted in a shortfall to other countries of 7 per cent or more. Furthermore, the UK imported some heavy crude oil from the Gulf since the North Sea produces light crude oil. Its main suppliers were Saudi Arabia and Iraq, although this could fluctuate. As importantly, the UK believed, quite erroneously, that from 1995 it would need to import light crude oil as North Sea supplies ran down.

While Saudi Arabia has tended to invest in the US, Kuwait and the lower Gulf states favour the UK. Inward investment from Kuwait in the 1980s accounted for between one-quarter and one-third of total foreign direct investment in Britain, roughly equivalent to the proportion from the US and from Europe, but more than Japan.[59] Middle Eastern banks in London 'made an important contribution to Britain's invisible income'.[60] However, the Gulf became even more important in terms of trade in the 1980s. While it has been argued that the focus of Britain's economic activity has shifted away from the Third World towards Europe and the US,[61] in contrast to these latter markets in the 1980s, the UK enjoyed a trade surplus with the Gulf – some £3 billion in 1987 – overturning the deficit of the 1970s.[62]

Given these interests, it is true to say that the stability of the region was of some importance to the UK: however, its ability to ensure that stability was highly questionable. At best, the UK's efforts must be seen as a *contribution* to that of the US, primarily because it lacks the necessary military forces: the Armilla patrol, for example, was only important in terms of the Gulf shipping it protected and in 'reinforc(ing) the British diplomatic presence . . . and the perception of Britain as a friend of the Gulf rulers'.[63] Even that limited effort – two frigates and a destroyer with support ships – was 'an expensive commitment',[64] and opposed by the Royal Navy on grounds of cost.[65] Britain's military instruments for maintaining the status quo lay, in Lady Thatcher's words, in the 'supply (of) equipment, training and advice'.[66]

What contribution arms exports made to the stability of the Gulf is a matter of dispute. On the one hand, it might be argued that arms supplies substitute for a military presence – a consideration which has been relevant to both the UK and the US in a region where the stationing of Western forces has domestic political consequences for the host countries. On the other hand, apart from the general arguments against weapons supply in terms of regional arms races and so on, one could argue that arms supplies do nothing to resolve the *particular* security problems of the Gulf. Kelly, for example, argued that both Iran and Saudi Arabia, with their size and territorial claims upon the smaller Gulf states, represented 'the principal threat to the Gulf's tranquillity'.[67] In this case, the supply of British arms to the two countries has done nothing for Gulf regional stability. Moreover, Kelly went on to argue that:

> If the Gulf states had been equipped according to their real defensive needs, as they were in the past, they would now possess little more than small arms, light artillery, armoured cars, helicopters, naval patrol craft and other military equipment of an equally modest order. What they have accumulated instead in the way of powerful and costly armaments will not serve to protect them from a major power bent on their subjection.[68]

Iraq's invasion of Kuwait subsequently underlined his point.

Arming the major powers of the Gulf has been justified in terms of the global struggle with the Soviet Union. By supplying weapons to Saudi Arabia and Iran, and in the 1980s to Iraq, the West not only demonstrated its interest in the region but prevented the Eastern bloc from gaining influence through its own arms exports. Since other means were available to demonstrate the importance of the Gulf to the West, the Western conception of stability had best be seen as meaning the maintenance of existing pro-capitalist states, of which the key regime was, and is, Saudi Arabia, and the containment of socialist or, in Iran's case, anti-Western, powers. It is here that arms supplies have had their real impact. First, defence imports served to buttress internal security; and second, they provided influence in that, as Kelly argued, Western governments were 'catering' to the Gulf states' 'obsession with military might'.[69] The cynicism behind Western arms supply to the region was recorded in a Foreign Office paper anticipating defence exports after the ceasefire in the Iran–Iraq War:

> We can take advantage of the fear of arms sales to Iran and Iraq by others to keep up the momentum of our defence sales to the Gulf States while moving gradually into the markets of Iran and Iraq.[70]

In short, arms exports to the Gulf had little to do with regional security but everything to do with Western economic interests. Although outsold there by the Soviet Union, the US and France, the Gulf represented Britain's best market for arms.[71] However, for the UK, defence sales to the region were both an end in themselves and a means of access to the wider market. They did not represent a substitute for military force in the discharge of its so-called 'historic' role in the region.

If that were the case, sales to Saudi Arabia and Iran would have been precluded since the UK's historic links were with the smaller states of the lower Gulf. Arms sales substituted for military force only in the sense of being a means of influence. However, one should be clear about the nature of that influence. Sir Anthony Parsons, formerly British Ambassador to Iran and the UN, argued, with respect to the UK and the Middle East, that 'What one cannot expect is for us to exercise what the Americans call "leverage" over this party or that.'[72]

From the Gulf experience, the influence which accrues from arms sales is altogether a more subtle phenomenon. The recipient viewed arms supplies as a litmus test of the state of relations between purchaser and exporter. In turn, shares in the general market were, to quote Sir Anthony Parsons, speaking of Iran, 'determined to some extent by the political relationship'.[73] Alan Clark supplied the complete list of determinants as a 'mix of influence and habit and corruption plus, of course, the quality of the goods that you supply'.[74] This state of affairs is true for all arms suppliers and is what gives rise in part to the phenomenon of 'reverse influence'. Countries which have the markets, and the wit to exploit them, are able to influence suppliers – at least to provide them with the arms and technology they require.

It has been argued that it was the lure of a defence contract, in excess of £100 million, with Iran, as well as substantial military deals with Saudi Arabia, rather than economic exigency, which persuaded the Heath Government not to reverse the decision to withdraw from the Gulf, and to assist the two countries' respective territorial claims in the region.[75] Since Britain retains little power in the Gulf today, defence contracts and civil trade are used in more modest ways by the recipients. For example, as Stanley and Pearton argued: 'Britain, out of consideration for the economic importance of her relations with the Arab countries has, since 1956, been consistently inhibited in selling arms to the Israelis.'[76]

An informal veto over arms sales to Iran and Iraq was exercised by Saudi Arabia, particularly after the first part of the *Al Yamamah* defence contract was signed with Britain in 1986. In 1988, for example, the MODWG, in operating the Guidelines, was asked to bear in mind 'the fact that Saudi sales are more valuable than sales to Iran'.[77] An FCO document argued that a gradual relaxation of the Guidelines after the ceasefire should be 'sensitive to the concerns of our best *customer* in the Middle East, Saudi Arabia'.[78] (emphasis added). Again, the export of small boats to Iran was blocked because, to quote an FCO memorandum, written by a senior official, 'The real argument against supply is that the Saudis and perhaps the Americans will object and the Tornado financing package would be jeopardised.'[79]

Paradoxically, arms sales had to be restricted in order to maintain them. But the UK had competing interests in Iran: the Guidelines provided the mechanism for restricting sales to maintain access to *all* markets in the Gulf.

Both Iran and Iraq made clear their willingness to use trade as a lever to acquire defence goods. This is not to suggest that the UK was loathe to supply military equipment as such: it merely serves to explain the balance that was sought via the Guidelines as well as the continuing trade with Iran in the face of US and Arab pressure. As a DTI memorandum said of Iran and Iraq: 'Both are state trading companies and have the means to steer civil contracts elsewhere if we are un-cooperative . . . the Iranians have made this clear.'[80]

Since all of the Gulf states are ruled by small élites or families, the relationship between defence and civil trade can be assumed to be similarly close. For example, after the first *Al Yamamah* deal, British civil exports to Saudi Arabia rose. The FCO argued that 'Tornado sales made a major contribution but should not be allowed to obscure a strong UK civil performance.'[81]

Iran, prior to the Revolution, had been a long-standing and important export market for the United Kingdom. In March 1979, Chase Manhattan Bank estimated that the country would lose £500 million in export sales to Iran in that year causing a loss of three-tenths of a percentage point from its 2 per cent growth rate.[82] Iran was the United Kingdom's biggest export market for arms. While the ECGD had paid claims for losses by UK firms of £100 million by 1981, some equipment, already paid for by the Shah, remained in the UK.

When the Iran–Iraq War began, Iran was already the subject of sanctions. However, negotiations to resolve the pre-Revolution contracts were continuing. In 1980, the UK was faced therefore with two countries, oil rich and industrializing, both of which were perceived to be of potential commercial importance. Iran had been the more important of the two, but the loss of the Iranian market had made the Iraqi market more significant than it might otherwise have been. The tilt towards Iraq in the early stages of the war reflects this perspective. The tilt was also the result of the new opportunities offered by the Iraqi *infitah*, the economic liberalization begun in 1979,[83] and the lifting, in the same year, of Iraq's partial embargo of British goods.[84] Sir Stephen Egerton, the Under-Secretary for the Middle East, confessed that the tilt towards Iraq was 'for commercial reasons'.[85]

As the Minute of a meeting in January 1981 of the Overseas and Defence Committee of the Cabinet has it:

> Ministers . . . agreed that though lethal arms and ammunition would not be supplied to either side, every opportunity should be taken to exploit Iraq's potentialities as a promising market for the sale of defence equipment and to this end 'lethal items' should be interpreted in the narrowest possible sense and the obligations of neutrality as flexibly as possible.[86]

The combination of commercial opportunity in Iraq coupled with the lack of orders because of poor relations with Iran[87] at this point explains the so-called 'tilt', at least in what was officially supplied. A contributory factor might have been that short-term credit only became available to Iran from 1982. The figures in the note at Table 4 show that all trade with Iran was down in the first few years of the war.

Whereas post-revolutionary Iran, in the words of a former Ambassador to Iraq, was 'a threat to stability in the area and consequently to the maintenance of British interests in the Gulf',[88] the FCO feared that greater restrictions upon the defence trade might mean that Iran would 'retaliate against general trade'.[89] At least until 1987, the importance of that general trade compared to that of Iraq is evident from the figures in Tables 4 and 5.

|  | 1983 | 1984 | 1985 | 1986 | 1987* |
|---|---|---|---|---|---|
| *Iraq* |  |  |  |  |  |
| Imports | 30,334 | 69,047 | 44,125 | 66,129 | 33,871 |
| Exports | 399,920 | 343,120 | 444,749 | 443,890 | 271,655 |
| *Iran* |  |  |  |  |  |
| Imports | 100,593 | 368,572 | 63,317 | 100,303 | 187,572 |
| Exports | 630,683 | 703,384 | 525,589 | 399,373 | 307,853 |

£ thousand

Source: Foreign Affairs Committee, *The Iran–Iraq Conflict*, Minutes of Evidence, Session 1987–88, House of Commons Papers 279–vii, 20 April 1988, p. 110.

*Years for which comparative figures available. Exports to Iraq in 1981 were £624 million; Iran, £403 million; and for 1982, £874 million for Iraq and £334 million for Iran.[90]

Table 4: UK exports and imports: Iran and Iraq

|       | 1983 | 1984 | 1985 | 1986 | 1987 |
|-------|------|------|------|------|------|
| Iraq: | 369,586 | 274,073 | 400,624 | 377,761 | 237,784 |
| Iran: | 530,090 | 334,812 | 462,272 | 299,070 | 120,281 |

Table 5: Trade surplus: Iran and Iraq

The relative increase in the trade surplus with Iraq might be explained by reference to the trade credits offered by the UK. In total, £1,140 million of medium-term credit was granted to Iraq between 1983 and 1988, £875 million of which was taken up. Short-term credit was available throughout the conflict[91] – as it was to Iran from 1982.[92] As Sir Richard Scott argued, the provision of 20 per cent of this credit for defence goods from 1984 was 'plainly inconsistent with a policy of even-handedness'.[93] However, one should again hesitate before attributing such provision to a 'tilt' in Iraq's favour. Iraq initially asked for credit for business already covered on a short-term basis by the ECGD. The UK agreed to avoid losses on existing contracts. Further credit was asked for by Iraq after the 1982 collapse in oil prices, and since British trade with Iraq was rising, the Government agreed.[94]

The fall in exports to Iran from 1986 is also attributable to its lower oil revenues such that the Government considered increasing credit in June 1988.[95] Medium-term credit was not available to Iran and is explicable, not by reference to any political preference, but by the outstanding claims on the ECGD from the pre-revolutionary contracts, and, possibly, by Khomeini's refusal to borrow abroad to finance the war. Nonetheless, two factors were important for the UK in respect of Iran. First, there remained the pre-revolutionary contracts and the related losses incurred by the ECGD. Second, was Iran's *potential* as a market for British goods as the only country in the Middle East to combine wealth and a large population.

The 'sole concern' of the second Guideline was Britain's 'past contractual obligations with Iran'.[96] Negotiations regarding them were strung out throughout the course of the war and beyond. In July 1990, the MOD argued:

Very difficult negotiations with the Iranians have begun on the pre-revolutionary contracts. If these go badly, and the Iranians force us to arbitration, it could cost us up to £200 million or more. The dropping of the Guidelines would be a very useful factor in the current negotiations.[97]

Although trade with Iran had been curtailed after the Revolution in 1979, the volume of trade which continued to exist remained significant. Of particular importance was Talbot's sale of cars in kit form to Iran. The company's profitability relied upon this deal until 1984 when Peugot bought the firm.[98] British Petroleum also had an important stake in Iranian oil,[99] although UK imports of Iranian oil were small in comparison to oil imports from other Middle Eastern states.[100]

The determination to maintain the trading relationship with Iran meant that Britain supplied, in the words of Sir Stephen Egerton, 'things . . . of really considerable seriousness in the war effort . . . like clutches for Chieftain tanks, that would have annoyed the Arab community very, very much'.[101] As SIPRI argued: 'Other things being equal, it is in the situation of armed conflict that the supplier-recipient relationship becomes closest.'[102]

Thus, at the first meeting of the MODWG on 13 December 1984, the record stated that:

Defence Commitments (ROW) said he had seen a reply from the Secretary of State, submission on tank engine spares for Iran, which stressed the need for officials to bear in mind the need not to prejudice civil trade while operating within the revised Guidelines.

This was a point, said Lt Col. Glazebrook, 'that was made to us frequently'.[103]

The threat to disrupt civil trade by both Iran and Iraq was not an empty one. After the UK both took a central role in the formulation of UN Security Council Resolution 598 and closed the Iranian Military Procurement Offices in 1987, Iran retaliated with a partial trade embargo.[104] Again, in 1988, following the UK's support for the American presence in the Gulf, Iran instituted a 'Buy British Last' campaign.[105] Noting that Iraq would become 'the second market in the Middle East', the Minister of State at the FCO went on to argue that 'weapons are what the regime uses as a touchstone of reliability in a trading partner'.[106] As such, Iraq introduced a partial trade ban following seizures of parts of the supergun and so-called 'nuclear triggers'.[107] A month later, Britain liberalized the restrictions upon defence sales to Iraq.

Iraq, like Iran, might have been increasingly able to exercise pressure upon the UK because of the credit extended. The UK was anxious to retain its 'favoured trading partner' status in the latter part of the 1980s in order that Iraq continue, in the face of some £40 billion of external debt, to repay its loans to the UK. By 1989, Iraq was in repayment arrears to the UK of some £80 million. France was similarly in credit, although more obviously affected by the reverse influence which debt provides to the debtor.[108] Iraq's pressure upon the UK in terms of defence exports is evident in its strategy of preventing interference with its procurement network. It is possible that Iran was similarly engaged in procuring technology for proliferation and was similarly using its general market as a lever of influence.

## Summary

Analysts have missed the complexity of the UK's motives in supplying arms. This is attributable to the fact that British interests are rarely revealed in its policy declarations. The Guidelines are an exemplar in this respect. Ostensibly a policy for bringing the Iran–Iraq War to an end and for fulfilling the obligations of neutrality, they served, quite purposefully, to mask the UK's real interests. As an Assistant Under-Secretary in the FCO said of them in 1986:

> I believe the Guidelines should be regarded primarily as the set of criteria for use in defending against public and Parliamentary criticism, and criticism from the Americans and Saudis, whatever decision we take on the grounds of commercial and political interests.[109]

Accordingly, the decision-making process was constructed to allow those interests to be tested one against the other in the inter-departmental forum. The policy reflected, in the Foreign Secretary's words, 'the difficult balance that we have to strike between our Arab friends and the West's interests in Iran'.[110] The contest between Departments was about finding that balance; and the informal veto each had was the mechanism for ensuring that, as for any defence exports where the balance was uncertain, it was the political centre which made the final decision. In these senses, the Guidelines can be regarded more as a tool of policy than a statement of it. As the FCO privately noted, 'we cannot reflect . . . pragmatism in a statement of policy'.[111]

Foreign policy, as national and economic security, was the guiding force behind British arms exports. Also evident, although to a lesser extent, was the operation of foreign policy upon defence sales in the guise of a concern for Britain's 'position as a leading figure in the world'.[112] Whether this represents 'the hold which the "Great Power Syndrome" continue(s) to exercise over the British Government's collective imagination'[113] or simply a wish, for the sake of domestic public opinion, to appear so is unclear. The FCO seems divided among those who believe that Britain should maintain a reputation for responsibility and those who see its interests as lying on the more concrete foundations of trade. Certainly, in considering the revision of the Guidelines after the ceasefire, there are several references, in support of retaining them, to Britain's position as a Permanent Member of the Security Council. But this concern seems allied to domestic public opinion. An IDC paper argued:

> Neither public nor Parliamentary opinion would understand if the UK, as one of the prime movers at the UN in promoting a peace settlement, were to dismantle arms export restrictions before we could be reasonably certain that both sides were committed to peace.[114]

The Government withheld defence equipment when Iran and Iraq behaved in unacceptable ways, but the effects were in general in response to public opinion and short-lived with the exception of the Rushdie affair. However, the UK did seek influence through its arms exports. The references to it are numerous. As an internal MOD memorandum noted:

> Apart from their commercial importance, defence exports, or the training or loan services systems that generally accompany them, can be a powerful lever of influence and are therefore important in defence as well as foreign policy terms.[115]

While one would not want to rule out the seeking of influence for reasons other than trade simply because the evidence is in scant supply – for example the access to Iran and Iraq which the defence trade gave Britain might have been a useful factor in its relationship with the US as well as in serving more general Western interests – the overwhelming preoccupation seems to have been with trade. That preoccupation was with the overall market, not specifically with the defence trade, save in the case of the enormous *Al Yamamah* deal. Defence sales were not simply an end in themselves nor merely a means of preserving the defence industrial base. Indeed there are no references to the latter in any of the documents, although that might be because it was a given for officials and Ministers.

The British, as David Mellor pointed out, 'are exporters', and the Middle East was of importance because, unlike the European market, Britain had a trade surplus there.[116] This should be seen against the backdrop of a continual balance of payments problem. While arms might have brought influence, these economic factors meant that Hegel's dialectic of master and servant obtained in the UK's relationships with the Gulf. Arms brought influence in that they provided access to the general market. The importance of that general market then became a tool of influence for the *recipient*. The trade in arms and technology became more important than other factors in the relationship. The political process became, as Greenaway put it, 'the servant of the economic process'.[117] The Minister of State at the FCO writing in 1989 underlined the point:

> Iraq's regime is one of the most vicious in the world. They are aggressive, use torture and repression, have used chemical weapons widely against Iran and the Kurds and their diplomats have behaved intolerably in the UK. The judgement is, are we strong enough, as a trading nation, to spurn their market on the grounds of morality? . . . On balance I judge that we are not.[118]

While such influence was not problematical for the UK where it wished to supply defence equipment in any case, or to withhold it from third parties, it was where it might have wanted to withhold arms exports for reasons of diplomacy, i.e. to signal displeasure, or where it might have preferred not to supply certain equipment, for example technologies useful to the development of weapons of mass destruction. In short, reverse influence operated via civil trade where the surplus was in the supplier's favour. That civil trade was not substantial: Iraq by 1990, on similar trade figures to those which obtained throughout the 1980s, represented half of 1 per cent of UK exports.[119]

Although Britain might seem uniquely prone to this sort of recipient pressure because of its economic decline and its trading position, it should be remembered that this form of influence by Iran upon the US was noted as long ago as the 1970s. There is no reason to suppose that other governments, in competitive international and domestic environments, are not similarly influenced. This is not to say that such influence will always be effective. Like any form of influence, it has its limits. As in the case of the UK and Iran, a serious breach in diplomatic relations with attendant public outcry, will have an impact upon what defence goods are supplied. Similarly, in the case of Iraq, the valuable Hawk deal was blocked because the *balance* of economic and political interests was against it. Conventional arms exports decisions are not, for the UK, straightforward, as some analysts have implied, because, as a highly political act, they have consequences for other states besides the recipient. And the fact that such third parties have a ready mechanism for expressing their disapproval means that foreign policy considerations will always be uppermost in the decision-making process. Ironically, governments seem better able to hide the defence trade from their own populations than from each other.

Whether influence and trade were the main motives for the *totality* of the UK's arms export policy towards Iran and Iraq or whether more precise objectives were also involved – in terms of the outcome of the war – is the subject of the next chapter.

# 5

# Covert defence sales to Iran and Iraq and the Gulf War

## The grey market

The examination of the covert export of British arms to Iran and Iraq during the Gulf War is an investigation into what Klare called the 'grey market'.[1] As Karp noted, it is, like the black market, 'all but ignored in orthodox studies'.[2] Klare argued that:

> Sales of military and military-related goods that fall, for one reason or another, between black-market sales and open, legitimate transactions compose the 'grey market'. Such 'greyness' originates in the nature of the item being transferred, or in the character of the transaction, or both.[3]

An inquiry into the sale to Iran and Iraq of unambiguous military goods using conduits and false end use certification is concerned with the 'character of the transaction'. The investigation of British sales of dual-use technology is treated separately. This separation is made because, in the case of dual-use exports, issues of deception by the recipient and control by the supplier are of a different character. Moreover, the export of that dual-use technology which contributes to WMD programmes has different security implications for the supplier than the export of conventional arms and, therefore, the motives involved require careful and separate examination.

While the black market is 'anti-policy', the grey market represents 'policy in flux as governments experiment with new and risky diplomatic relationships'.[4] It is the contention here that, in spite of the prosecutions of small companies who exported arms to Iran and Iraq, the British Government knew and approved of the grey trade with the two countries.

The discovery of motive here is complicated by the fact that no hard and fast data exist about the totality of British covert exports, and by the related difficulty in assessing the contribution of those arms to the course of the war. In other words, it is uncertain whether the unofficial policy was directed at the war or at relationships with Iran and Iraq. But, before motive can be ascertained, it is necessary to evaluate the evidence of British grey market sales to Iran and Iraq and, more importantly, what knowledge the Government had of them.

While sales to Iran were in general judged to be beyond the Inquiry's scope, Sir Richard Scott found no evidence of government complicity in covert arms sales to Iraq, except in the case of Jordanian diversion. However, Sir Richard looked for direct evidence, but by its very nature, indications of connivance by government will be circumstantial. It is, therefore, the *weight* of that circumstantial evidence which should be assessed.

In trying to establish what knowledge the Government had of the diversion of British arms to Iran and Iraq, the premise here is that if information was in the public domain then the Government might reasonably be assumed to have acquired that same information – Departments subscribe to cuttings services. Proving government complicity in arms sales to Iran is not as straightforward as it is in the case of Iraq. However, there are certain facts with which wide-ranging intelligence agencies and military personnel can be expected to be acquainted, such as the use of false end user certification and conduits in the arms trade, as well as what are, and are not, reasonable imports of arms by certain countries. Thus, where export licences were requested for countries which did not have a reasonable need for equipment, or which were known to have acted as conduits or transshipment destinations, it is reasonable to expect suspicions to be raised. Greece was reported to be supplying Iran with arms in early 1984[5] and in 1986 the press covered the provision by both the Philippines and Yugoslavia of false end user certification.[6] Austria is referred to in an AWP minute from 1986 as a 'known supplier of Iran and Iraq'.[7]

Swedish and French Customs conducted investigations, from late 1984 and 1986 respectively, into cartels of European arms manufacturers supplying both Iran and Iraq. These inquiries ought also to have alerted the British Government, if they did not already know, to, not only the identity of the countries providing false end user documents and acting as conduits, including Singapore, but, more importantly, to the involvement of British companies.

Other evidence of the Government's knowledge of, and support for, 'grey market' arms exports to Iran and Iraq comes from both the nature and frequency of the allegations made by business people, and the behaviour of British companies. On the latter point, in the case of Mark Gutteridge, the Sales Manager, and Paul Henderson, the Managing Director, of Matrix Churchill, intelligence was, in effect, traded for the approval of export licences for Iraq. From Chris Cowley, the Project Manager for the supergun, as well as Dr Bayliss of Walter Somers, comes evidence that, for arms manufacturers, the support of the Government was crucial to their continued trade.[8] That support was retained through honesty. For example, when Astra acquired Pouderies Réunies de Belgique (PRB), it informed the MOD, within weeks, of PRB's existing propellant contracts. In fact, the Deputy Head of Defence Sales approached Gerald James and told him to report 'anything untoward' in PRB's contracts to the MOD immediately.[9] It would therefore be wrong to assume that Paul Henderson was unique in providing information to government.

## The diversion of British arms by Jordan

By far the strongest evidence relates to Iraq and, more especially, the sale of arms via Jordan. Before considering that particular case, the evidence that Saudi Arabia, Kuwait and Egypt acted as conduits for Iraq's military supplies should be noted. Iraq, because of its geographical position, had to rely upon supply routes through countries such as Saudi Arabia, Kuwait and Jordan, and this assisted the turning of the blind eye towards diversion by the British Government. As early as 1981, Saudi Arabia was identified as a diverter of arms to Iraq.[10] *Jane's Defence Weekly* reported that Kuwait was supplying Iraq with British Chieftain tanks in 1984,[11] and Kuwait was referred to as a known diverter of CW equipment to Iraq in a 1985 JIC paper.[12]

A DIS minute from 1984 refers to 'quite conclusive' evidence of links between Egypt and Iraq in the CW field,[13] and a report by the SIS in 1985, distributed in the FCO and the MOD, referred to Egyptian Military Attachés trying to obtain ammunition for Iraq from various countries including the UK.[14] In 1986, the UK received Israeli intelligence that Egypt 'had purchased substantial quantities of hydrogen fluoride on behalf of Iraq'.[15] The following year, the export to Egypt of extrusions, parts of the body of 122mm ground-to-ground rockets, were refused following DIS and security advice that these castings would be diverted to Iraq, and could be fitted with chemical warheads.[16] In 1987, Jordan, Saudi Arabia, Kuwait and Egypt were placed on a Ministry of Defence list of diverters of NBC goods.[17]

The issue of Jordan's role as a diverter of British military exports to Iraq was first raised in the House of Commons in December 1980 when Greville Janner asked the Secretary of State for Defence 'if he will ensure that any British arms sales to Jordan are conditional upon those arms not being resold to Iraq'. The reply represented no commitment to do so: 'Export licences are only granted for arms sales once we are satisfied that the request is authentic and the arms are destined for the exclusive use of the purchasing Government.'[18]

The first concrete evidence that the UK understood Jordan to be a conduit for arms supplies to Iraq dates from 1983 when Customs and Excise found that 'officials of the Iraqi and Jordanian Embassies were involved in producing fake end user certificates' for Sterling machines guns.[19] However, intelligence reports of Jordanian–Iraqi collusion on arms supplies also date from 1983.[20] The Army Department opposed the sale of a large quantity of 120mm APFSDS tank ammunition to Jordan on the grounds of diversion to Iraq and registered scepticism about the value of end user assurances.[21] The MOD wrote to the FCO in March 1984 to say, 'It has also been possible to identify sales of British ordnance to Iraq through third parties.' The letter gave examples.[22] An intelligence requirement for Jordan began in October 1985 and was extended each year.[23] A JIC report in December 1985 noted that:

Despite the controls on the sale of certain chemical precursors, Iraq has still managed to obtain supplies, either through intermediaries, such as Kuwait and Jordan, of from countries not imposing controls.[24]

In the following year there were at least four intelligence reports concerning Jordan's diversion of military equipment to Iraq. A Secret Intelligence Service report dated 14 March 1986 noted that 'the Iraqis have no problem over obtaining equipment thanks to the willingness of certain countries to act as a notional end user'. Jordan is referred to as one such country.[25] In succeeding years similar reports flowed in. Moreover, the concerns of the DIS and the AWP about Jordanian diversion of British arms can be documented throughout the 1980s.[26] The British Embassy in Amman wrote to the FCO in 1987 to draw attention to the coincidence of an increase in Britain's trade with Jordan and a decrease in trade with Iraq. The letter speculated that 'Jordan may be acting as a way round Iraq's difficulties in obtaining credit from Britain.'[27]

The knowledge that Jordan and other Arab states were supplying Iraq throughout the Gulf War was not confined to officials. David Mellor recalled a visit to the Gulf when, as Minister of State at the FCO, he travelled 'by land from Baghdad all along the front line down to Basra and into Kuwait and ... saw ... the flow of goods'.[28] A senior diplomat recalled 'those lovely roads' from Aqaba in Jordan to Baghdad and the fact that 'the Iraqis had a whole section of Aqaba port run by themselves ... called (the) "Iraq Ports Authority"'.[29]

The Gulf Arab states' support of Iraq was well known. Equally, according to the Jordanian Ambassador to the UK, 'it was no secret' that Jordan, among others, operated as a conduit for Western arms to Iraq.[30] While witnesses before the Scott Inquiry have insisted, to quote one of them, that 'whenever we are mopping up the Iraqis' kit in Kuwait, there (was) not a single piece of evidence against the Jordanians for diverting British equipment',[31] participants in the trade state that they saw military goods from the UK in Iraq's enclosure at Aqaba.[32]

At an early stage in the war, Iraq captured about 260 Chieftain (Shir) Main Battle Tanks (MBT) from Iran. Jordan had been supplied with the Khalid version of the Chieftain tank as well as Chieftain Armoured Recovery Vehicles (ARV). When considering an AWP application, in 1985, for Jordan for Chieftain spares, the Arms Working Party noted that 'it must be borne in mind ... that (Jordan) could be buying them on behalf of Iraq'. The application was cleared.[33] Another AWP application in 1986 for 10,000 spares for Chieftains was cleared in spite of Jordan being 'a known supplier of Iraq in the defence field'.[34] Further AWP applications and ELAs for – often specifically Chieftain as opposed to Khalid – spares were subsequently approved the following year, even though it was known that Iraq was 'seeking Chieftain spares'.[35] It should be noted that, although both Iran and Iraq could receive some spares for their respective tank fleets, they could not be supplied with spares for the guns and hulls. Jordan, however, could receive all MBT spare parts.

The IDC minutes of 11 August 1988 note that in a meeting with the British Ambassador and the HDES:

> Field Marshall Bin Shakir said that ... the Iraqis had asked Jordan 'to front for them' in the supply of certain military equipment especially spares for Chieftain ARVs and MBTs ... Field Marshal Bin Shakir ... had told the Iraqi Government that they were not prepared to do anything behind the back of the British Government. The most they were prepared to do was to relay the request for us to consider.

The IDC recommended to Ministers that they agree 'in principle' to the supply of automotive and miscellaneous spares 'direct to Iraq' and hull spares 'if the vehicles are repaired in Jordan'. Furthermore, the IDC minutes of 11 August 1988, as well as subsequent correspondence, noted that 'The UK supplied Iraq with 29 ARVs in about August 1981 under the Jordan Memorandum of Understanding (MOU)'[36] with about two years supply of spares. In July 1985 Iraq requested spares for its 29 ARVs.[37]

After the signing of the second MOU between Jordan and the UK on 19 September 1985 – a 'very nice little contract' worth £270 million[38] – the 'great majority' of ELAs under the agreement were approved 'automatically because they (were) part of the package'.[39] The original package had been agreed by the Assistant Chief of Defence Staff, Operational Requirements (Land) who established that the types and quantities of military equipment met the reasonable needs of the Jordanian Armed Forces.[40] Lt Col. Glazebrook, of the MOD's security staff, was asked by Sir Richard Scott to re-examine this assessment and found that, on the whole, the quantities did indeed met the reasonable needs of the Jordanian Armed Forces.[41] Furthermore, the MOU contained the 'standard clause', section nine, which prohibited disposal to third parties.[42] However, one should bear in mind three important points. Much of the equipment supplied was in use in both Jordan and Iraq. The package was expanded by some 15 per cent in February 1987.[43] As the Scott Report noted, the MOU did 'make it very difficult for AWP members to object to the grant of licences on the ground of the risk of diversion to Iraq'.[44]

The common equipment between Jordan and Iraq, a list of which was only supplied to the IDC in May 1990, included Chieftain tanks, surface-to-air missiles, guns, mortars, and armoured personnel carriers, and ammunition.[45] As the size and shape of the Jordanian package altered, Lt Col. Glazebrook 'had an uncomfortable feeling that there were things going on behind my back about which I was not aware'. The DESO claimed equipment was part of the package which was subsequently found not to be so.[46] Moreover, the AWP was put under pressure to 'give an instant clearance to additional items',[47] and Lt Col. Glazebrook, known for his assiduousness, was not given the list of extra equipment until three months after its original circulation. The list included thermal imaging equipment, laser range finders, tank transporters, fast patrol boats and cupolas for MBTs.[48]

Also facilitating the diversion of British defence equipment was the use of the OIEL for countries known to be fronting for Iraq. The OIEL authorizes a company to export a specified type of equipment in unspecified quantities. Companies were only obliged to keep records from 1990, with inspections beginning in 1991. As Sir Richard Scott remarked, the 'rapid removal' of Jordan from the lists of authorized destinations after the imposition of the UN embargo on Iraq 'tells the story'.[49]

Sir Richard Scott argued that Government policy on exports to Iraq was 'undermined by the use of Jordan as a diversionary route'. If, as witnesses claimed, that was 'a price worth paying' to maintain good relations with Jordan, then that decision should have been 'specifically put to and taken by Ministers'. Sir Richard saw no evidence that that was done.[50] But, on the one hand, there was an abundance of intelligence circulating within Government to show that Jordan, among other Gulf states, was acting as a conduit for Iraq; and, on the other, evidence that the UK actively connived in this diversion – viz Field Marshall Bin Shakir's request. A senior FCO official argued that, regarding Jordan, 'Whitehall as a whole was sensitive to this problem as is evidenced by the intelligence requirement.'[51] Mark Higson, a former Iraq Desk officer commented: 'Indeed, even during my time in the British Embassy in Kuwait (1983–86), we knew Jordan through Aqaba was being used for imports of hardware from the UK, which (were) going on to Iraq.'[52]

Alan Clark went still further: 'More than half the material purchased by Iraq was actually consigned to Jordan . . . there was a tendency . . . for the trickier items to go to Jordan.' In any case, he argued, 'false end user information almost (has the) status of common practice in the Middle East'.[53]

## The European cartels

Less substantive is the evidence of government knowledge of the arms manufacturers' cartels which supplied both Iran and Iraq. Swedish Customs began investigating the propellant and powder cartel in 1984. Its 7,000-page report published in 1987 was summarized by the Swedish Peace and Arbitration Society.[54] Bofors (which includes Nobel-Kemi), co-ordinated various European manufacturers in an operation to supply Iran and Iraq during the Gulf War. The supply began in 1981 and ended six years later. The purpose of the propellant cartel was to agree co-production prices and the division of markets, not just in respect of Iran and Iraq. Orders for explosives from the belligerents were so large that a mechanism for sharing contracts was necessary. The cartel used an existing organization, the European Association for the Study of Safety Problems in the Production of Propellant Powder (EASSP), as a cover for the co-ordination of orders and shipping arrangements. Scandinavian Commodity (SC), owned by Karl Erik Schmitz, a businessman who acted as an agent for Iran, brought the first Iranian propellant contract to the cartel in 1983. It was for 105mm and 155mm ammunition and valued at $155 million.

The cartel used false end user certificates, principally arranged by PRB, a company that was bought in 1989 by the British firm Astra. Countries such as Kenya, Zambia, Yugoslavia, Rumania, Czechoslovakia and Singapore were used in this regard. Shipping documents at Appendix A show two identical manifests, one carrying the true destination, Bandar Abbas in Iran, and the other a false one, Lavrion. That the cargo was destined for Bandar Abbas can be seen from the bill of lading and the manifest when the ship docked at Zeebrugge from Ridham – an obscure port on the River Swale – en route to Spain and Greece before arrival in Iran. Circuitous banking arrangements were used to hide Iran's purchasing. Documents show that Bank Melli of Iran issued a Letter of Credit paid to a Western bank. The bank then paid the manufacturer. There is no commercial necessity in using a second bank in this way.[55]

ICI's subsidiary, the Nobel Explosives Company (NEC) and the Royal Ordnance Factory (ROF) were involved in the cartel. At a cartel meeting on 1 February 1984, the ROF were recorded as producing the explosive RDX, and attempting to get export licences for Milan. RDX, according to the Institute of Explosive Research in Washington, has a long shelf life and is favoured for sabotage operations.[56] While the ROF was not a member of the cartel, it was drawn into its operations. Bofors visited the ROF in 1984, and a cartel document shows the ROF as having, in 1981, a 2,000 tonnes annual share of the cartel's Iranian business.

At a cartel meeting in Paris on 10 October 1984, the minutes recorded that Guy Chevallier, head of the French explosives company SNPE and Secretary General of EASSP, 'had spoken to Truman who has promised to co-operate with the marketing'.[57] Trevor Truman was the director of the ROF's explosives division. The involvement of the ROF can also be seen from the shipping documents at Appendix A.

ICI, on behalf of the NEC, admit involvement in the cartel, but claim it was 'unwitting'.[58] The NEC, in contrast to the ROF, was a member of the cartel and aware, according to Swedish Customs, that the ultimate end user of the explosives which Bofors asked it to produce was Iran.[59] In any case, the size of the orders would tip off an experienced company that the purchaser was at war. The NEC is referred to on some twenty-six of the 200 pages of the cartel's minutes. In 1984, ICI forbade its subsidiary to attend meetings, prompted by accusations of price-fixing, but the cartel agreed to maintain contact with the company.

Thus far, the evidence has related to Iran but, as already noted, the cartel supplied both Iran and Iraq. The minutes avoided mention of the countries, but there are two, unusually careless, references to Iraq.

The Scott Inquiry did not investigate the cartels because there was no evidence to show government knowledge.[60] However, that knowledge of the existence and nature of the cartel can be deduced from the following. First, Swedish Customs began investigating the cartel in late 1984. It seems probable that, having discovered the involvement of British companies, Swedish Customs, or the West Germans who first tipped them off, would have contacted their counterparts in the UK. British Customs said they received a report in 1986, but made no reference to any communication before that date.[61] Second, from the shipping documents, one can see the involvement of the ROF in June 1986. On 2 January 1985, the MOD became the sole shareholder in the ROF before its privatization in 1987. According to the Memorandum of Understanding, the company was not permitted to enter new deals without the agreement of the Secretary of State for Defence. In any case, there would be no incentive for a nationalized industry to conduct business of which the Government would not approve. The MOD's response, when these facts were put to it, is revealing in not addressing the connection between the ROF and the cartel: 'We are not aware of the cartel. Royal Ordnance had some foreign sales which would have been cleared by the normal procedures.'[62]

The Government admitted in 1995 that the ROF exported direct to companies in Singapore which arranged shipment to Iran.[63] The ROF also exported ammunition to Austria, Portugal, Spain, Yugoslavia, Egypt, Saudi Arabia and Jordan in the 1980s, although Lt Col. Glazebrook, on examining the ELAs for the Inquiry, found no obvious reason to assume the end user was other than that stated.[64]

Both ICI and the ROF were investigated by British Customs in 1987, the result of reports from Swedish and Belgian authorities, but no evidence was found that the companies knew of the ultimate end users of their exports.

Also supplying both Iran and Iraq during the Gulf War was a British company called Allivane International Ltd. The company began supplying Iran in 1983 and Iraq in 1985. Allivane supplied Space Research Corporation (SRC), Dr Gerald Bull's company that provided long-range guns to Iraq, and had links with Carlos Cardoen, the Chilean arms supplier to Iraq. The Scott Report concluded that 'substantial quantities of ammunition were being exported from the United Kingdom to Iran' by Allivane.[65] Allivane set up a ghost company, Allivane International Group, in 1986, for the purpose of shipping arms to Iran and Iraq. While no such company is registered with Companies House,[66] export licences were issued to this ghost company.

Allivane was first drawn into the Iranian market by the French company, Luchaire. When, in 1986, Luchaire was exposed as being involved in supplying arms to Iran, Allivane is said to have taken over the co-ordination of the network involving ten contracts which met Tehran's orders for shells[67] – some 1.5 million of them between 1983 and 1988. Just as Luchaire's Chief Executive testified that he had the consent of Charles Hernu, the French Defence Minister, so Allivane executives are alleged to have had regular briefings with senior MOD officials, mostly from the DESO. In his testimony, Luchaire's transport manager alleged that Mario Appiano, a senior executive, travelled to London after each shipment involving Allivane. He had meetings with bankers, to execute the Iranian Letters of Credit, and with both Iranian and British officials.[68]

In April 1987, the SIS issued a report, circulated to the FCO and the DIS, which identified Allivane as supplying charges and fuses for a Portuguese company, Spel, for shipment to Iran.[69] Following information from the FCO, an Allivane executive and Dutch Customs, an investigation was instituted by Customs and Excise. The officer responsible, summing up his work in a report in 1988, noted 'Allivane's links with a cartel involving Dutch, French, Italian companies and other foreign agencies'.[70] In June 1988, Lt Col. Glazebrook noted Allivane's 'bad reputation' for diversion:[71] from October 1987 to November 1989, the REU had 'numerous discussions' about Allivane's supply of munitions to both Iran and Iraq.[72] This evidence of government knowledge sits uneasily with the MOD's assertion that it had no knowledge of the cartel.

An Italian company, Consar, and its Hong Kong subsidiary Sea Consar, as well as Sea-Erber, also an Italian company, the Dutch manufacturer, Muiden Chemie, and the Portuguese Spel company were involved in the Allivane network. ICI's NEC, the ROF and British Steel confirmed that they supplied Allivane with components and explosives, but denied knowing that the destination was Iran.[73] The destinations of the goods on Allivane's export licences include the Philippines, Singapore and Austria, all known routes for diversion to Iran. Allivane was supplying large quantities of a kind of mortar fuse to Singapore which was not used by the Singaporean Armed Forces but was used by the Iranians.[74]

Allivane supplied Iraq with arms using conduits such as Egypt, Saudi Arabia and Austria. Sir Richard Scott concluded that there was no evidence that the Government knew of this diversion.[75] However, the Inquiry discovered that Lt Col. Glazebrook had seen only seven of twenty-six Allivane ELAs for Austria, Egypt, Spain, Portugal, Saudi Arabia and Jordan. The ones he did not see: 'covered potential exports of large quantities of ammunition components to a variety of countries which were suspected, and in some cases known, to have diverted equipment to Iran and Iraq'.[76]

Furthermore, in 1988, in the midst of fulfilling a large contract for Saudi Arabia, the company's financial problems caused an interruption to the shipment. MOD officials on the REU 'believed it possible that the ultimate end user would be Iraq'.[77] Nonetheless, the shipment went ahead with the MOD involved in ensuring that arrangements were met, ordering, for example, the haulier, Frank Machon, to carry three times the legal load on his lorries and arranging storage at MOD facilities. The then Secretary of State for Defence said he 'probably did' make such arrangements.[78]

## BMARC

Sir Richard Scott concluded that there was no evidence of the Government's knowledge of exports to Iraq by the British Manufacture and Research Company (BMARC), even though the report noted that 'A number of files and documents that the Inquiry wished to inspect were missing.'[79] The company's exports to Iran were considered to fall outside the Inquiry's remit.

BMARC made naval guns and was owned by a Swiss company, Oerlikon Buhrle, until BMARC was bought by the British firm, Astra, in 1988. In 1986, Oerlikon had contracted to supply Iran with 140 GAM-B01 20mm naval guns worth £15 million. In all, 220 guns were supplied between 1986 and 1990 with spares and ammunition. They were to be shipped in parts for reassembly in Singapore and onward shipment to Iran via Pakistan. In the House of Commons in 1995, the President of the Board of Trade admitted that in 1986 the SIS knew that:

Iran had concluded a contract with Oerlikon for the supply of weaponry and ammunition. The intelligence picture developed in 1987 when it was revealed that naval guns made by Oerlikon had been offered to Iran by a company in Singapore. In July and September 1988 two intelligence reports rounded out the picture by referring to naval guns and ammunition being supplied by Oerlikon through Singapore to Iran.[80]

In fact, *Jane's Defence Weekly* reported, on 3 October 1987, that Iran had received Oerlikon twin 35mm towed anti-aircraft guns and their associated Skyguard fire control systems.[81]

According to the DTI's evidence to the Trade and Industry Committee investigating the affair, the 1988 intelligence was circulated to the DTI but, according to the Government, since there was no system for alerting officials to the fact that BMARC was a regular applicant for licences to Singapore or that Oerlikon was named as an agent in most of those applications, the connection went unmade. Weaknesses in the watch list system prevented officials from making the connection when further applications were received.[82] However, in all, Oerlikon/BMARC applied for forty export licences to cover the contract; and Singapore, a known conduit for illicit arms sales, had twelve Swift class coastal patrol craft each equipped with a single 20mm Oerlikon gun, according to the 1987-88 edition of *Jane's Fighting Ships*.[83] Oerlikon/BMARC was a known supplier of Iran. On 1 April 1984 the British press covered the training of ten Iranian officers, complete with picture, on the Skyguard fire control system which links twin 35mm anti-aircraft guns. Since Iran at the time had only older Oerlikon guns, and the Foreign Office approved the training, it can be assumed that the Government must have been aware of an order that had, or was to have, taken place. Local residents told *The Observer* that the officers had been at the BMARC base at Faldingworth since January and Iranians had been there before.[84]

BMARC, according to its Chairman, Gerald James (also the Chairman of Astra), was supplying both Iran and Iraq. Among the secret orders Gerald James discovered on taking over the company were: 210mm booster pellets for SRC's gun contracts in Iraq; 20mm and 30mm ammunition for Jordan; 2,250 tonnes of 155mm propellant for Allivane for Iraq; and Skyguard for various countries in the Middle East, including Iraq.[85] The Scott Report was able to substantiate the first, but not the last allegation.[86] The Company Secretary asserted that, 'The fact is the Government and the secret services knew all along what we were doing and could have stopped us at any time.'[87] Gerald James stated that the BMARC employee supervising the sale of naval guns to Iran was Major General Donald Isles. Since, argued Mr James, the General was a former Director of Weapons at the MOD, he would not have allowed the sale to proceed had the Government not known and approved it.[88] The General has denied knowing the true destination of BMARC's naval guns, as has the company's Managing Director. In addition to both the former Chairman and Company Secretary of Astra, the Chief Executive of Astra and an accountant with Stoy Hayward who worked on the acquisition of BMARC have asserted that the Government knew of the exports to Iran and Iraq.[89]

Since Sir Peter Levene, of the MOD's Procurement Executive, encouraged Astra to buy BMARC in 1987 in order to set up a competitor to the newly privatized ROF,[90] it is reasonable to assume that the DTI knew the connection between BMARC and Oerlikon. It is hardly credible, given the existence of intelligence on the one hand and the excessive quantities for the Singaporean Navy on the other, that the Government did not know about this particular British sale of arms to Iran. Alan Clark has admitted that Iran as well as Iraq 'got a certain amount' of lethal arms.[91]

Most of the companies thus far mentioned had many of the exports financed, and in some cases arranged, by either the Bank of Credit and Commerce International (BCCI) or the Banca Nazionale del Lavoro-Atlanta (BNL). The former was much more than a bank. As one of its operatives asserted:

> They could handle anything. Brokering, financing, Letters of Credit, false end user certificates, shipping, spare parts, training and personnel. You could order a bomb, a plane to deliver it and somebody to drop it.[92]

Like BCCI, BNL was underwriting arms sales to both protagonists in the Iran–Iraq War. Prosecutors in Venice indicted senior BNL officers and export licensing officers. The indictment identifies Allivane as supplying millions of fuses for artillery shells to Iran.[93] In 1993 the US District Court in Atlanta reviewed the BNL evidence before sentencing bank officials. The evidence included: National Security Agency reports; CIA documents; State Department memoranda ('the black book'); National Security Council reports and memoranda; Defence Intelligence Agency cables and information; the reports of the Italian Senate Commission and their evidence; the CIA report of the investigation of its handling of the BNL affair; and the Senate Select Committee on Intelligence Staff report on the involvement of the US intelligence agencies in the BNL affair. The Court concluded that:

> the defendant employees of BNL-Atlanta with their personal agendas and paltry awards were pawns or bit players in a far larger and wider-ranging sophisticated conspiracy that involved BNL-Rome and possibly large American and foreign corporations, and the governments of the US, England, Italy and Iraq.[94]

## The Iranian Military Procurement Offices

The final piece of evidence regarding illicit British arms sales during the Gulf War concerns the Iranian Military Procurement Offices (IMPOS) in Victoria, London. Although the IMPOS was officially closed by the British Government in September 1987, press reports subsequently alleged that it continued to operate, albeit with a reduced staff.[95] Described as 'the nerve centre of Iran's non-communist weapons purchases worldwide',[96] the IMPOS was allowed to continue by the British Government throughout most, if not all, of the Iran–Iraq War because, it was argued, there was no legal basis upon which to close it – neither the arrangement of arms deals nor the trans-shipment of weapons via the UK was illegal. Nearer to the truth, government sources argued that, despite persistent US pressure to close the IMPOS,[97] the security services preferred to keep their easy surveillance on an openly conducted arms buying network.[98] The London base suited the Iranians for three reasons: they sought spares for a mostly Western arms inventory; London represented a major financial capital where funds could be raised with ease; and there were large numbers of private arms dealers situated in the city.[99]

In terms of the Government's knowledge of illicit arms sales to Iran by British companies, the case of the IMPOS is instructive. Since the intelligence agencies could monitor most of Iran's arms buying via surveillance of the IMPOS, it can be deduced that the Government knew the identities of companies, if not contracts.

## Control mechanisms

Given that all of the companies thus far discussed were in possession of export licences, there remains the possibility that illicit British arms sales to Iran and Iraq represented no more than a failure of the regulatory machinery. Such a conclusion would have to be premised on the failure, in case after case, to connect intelligence or media reports with export licence applications. The responsibility for ensuring a system for detecting diversion to Iran and Iraq lay with Ministers: it was they who gave the IDC its remit. The argument could be made that the loophole was not brought to the attention of Ministers by officials; however, it remains to be seen what possible motive officials might have for such an omission. Sir Richard Scott noted that the speed with which measures were taken to prevent Jordanian diversion following the invasion of Kuwait made a telling contrast with official inactivity beforehand. No reasonable explanation suggests itself except that, both before and after 1990, officials were responding to political demands, whether tacitly or explicitly expressed.

The IDC was to consider only those applications specifically for Iran and Iraq. In September 1986, the IDC considered putting end user, or no transfer, clauses for all sales worldwide, thus demonstrating an appreciation of the spread of countries acting as conduits for arms supplies to the belligerents. However, this was considered to be 'too heavy handed'. Instead, 'Officials should ensure that the possibility of diversion to Iran or Iraq is taken into account in deciding on export licences for third countries.'[100]

Recommendations from a fairly junior official committee required enacting by a senior civil servant, in this case Sir David Miers, the Assistant Under-Secretary supervising the Middle East (MED) and Near East and North Africa (NENAD) Departments of the FCO. He did not remember 'doing anything to give specific effect to this particular recommendation', arguing that the FCO would look to the MOD to raise issues of diversion.[101] In the MOD, the Chair of the MODWG had ruled that the Guidelines did not cover the issue of diversion. While the security advisers and the DIS had argued that conduit countries should be considered in the operation of the restrictions on arms supplies, especially in regard to Egypt and Jordan, the DESO had opposed such an extension of the MODWG's remit.[102]

No system was introduced to prevent the diversion of British arms to Iran and Iraq. Those officials who had to apply the Guidelines did so with close, albeit subjective, attention. Questions of diversion were left to those who were not charged with the task of vetting potential defence exports to Iran and Iraq. It would not be reasonable to expect them to apply any criteria beyond the normal considerations of national security, commercial interest and so on, especially if they understood the implications of the IDC's remit.

Junior officials tried to cope with the issue of diversion while senior civil servants prevented the necessary institutional mechanisms from being put in place. Only two explanations suggest themselves for the conspiracy to allow goods to slip through the controls: either senior civil servants were designing their very own foreign policy; or they were aware, in a way that junior staff were not, of the political desires of Ministers. Thus, the conclusion must be that diversion to Iran and Iraq was irrelevant to the Government. As Sir David Miers put it with some candour:

> We have to recognise . . . that the importance of this policy was partly, in fact, in being seen to be operating a policy towards the belligerents. It was a political message rather than one that was actually going to affect the war . . . The important thing from our point of view was to be seen to be having a policy . . . If a consignment succeeded in getting around that policy . . . it was not actually going to affect the war very much. Nor was it going to affect the integrity of our policy, unless we connived at it.[103]

Possible explanations for the Government's toleration of 'grey market' exports to Iran and Iraq include the desire for trade and the quest for influence. In support of these objectives was the need for intelligence. A third explanation is that the UK sought, alone or in concert with other countries, to control the outcome of the Iran–Iraq War. This entails an assessment of the effect, or rather the *intended* effect, of arms supplies (and non-supplies) on the two belligerents.

## Covert sales as an extension of official policy

As well as asking what British interests might have been served in the illicit arms trade with Iran and Iraq, it is worth asking the question: What interests would have been *damaged* by participation in the 'grey market'? In so doing, some light is shed on the plausibility of allegations about the Government's complicity in covert arms supply. The export of 'lethal' arms would only have been inconsistent with official policy had the Guidelines been a response to either the obligations of neutrality or morality. Certainly, it would not have harmed either the pursuit of influence or commerce, unless, as Sir David Miers pointed out, either Iran or Iraq saw deliberate British connivance in the diversion of arms to the other party. The distinction between 'lethal' and 'non-lethal' was, at best, subjective and, at worst, meaningless: 'Anyone who has served in a military unit will know that it is more important to have a working radio than a full complement of small arms.'[104] For example, the UK sanctioned the sale of spare parts for *inter alia* tanks, and aircraft. Such official sales might also be said to constitute 'grey market' sales.

A few examples, although there are many, serve to make the point. The examples chosen are Iranian since, in terms of illicit sales, it is there that the evidence of government knowledge is circumstantial. The first concerns the supply of Rolls Royce Spey engines for Iran's F4 Phantom fighter aircraft in 1984. It was partly this export or, more precisely, the US reaction to it, that led to the formulation of the Guidelines. Washington argued that such supplies were 'partly responsible for keeping the Iranian air force flying'.[105]

The supply, in 1986, of parts for Boeing 707s for Iran was permitted, according to the IDC minutes, 'in line with our policy on civilian aircraft spares'.[106] However, at a later IDC meeting, it was noted that 'the only Boeing 707s operating in Iran are used . . . as airborne tankers for refuelling fighter aircraft, which might pose a threat to Armilla'.[107] In 1986, *Jane's Defence Weekly* noted that Iran's tanker force of Boeing 707s suffered 90 per cent unserviceability.[108]

In a similar vein, spares for ejector seats were permitted for Iran's F5 fighter aircraft 'on humanitarian grounds' (as for both Iranian and Iraqi aircraft).[109] The RAF security advisers had consistently argued that, if spares were not provided for ejector seats, 'the aeroplane would never have taken off in the first instance and would not have been available for lethal missions before it was actually attacked and crashed itself'.[110] Arguably then, exports of so-called non-lethal military equipment had much more importance to the recipient than so-called 'lethal' arms; and there might be more lethality in their effects, if not in their apparent qualities.

Since the Guidelines, while seeming to represent a restrictive approach, allowed some militarily very useful supplies to both Iran and Iraq, illicit sales might be seen as no more than an extension of those exports. All that was important, in policy terms, was that the latter should have no official approval attached to them. Just as the

Guidelines were formulated to maintain relationships with the two belligerents, and their supporters, British illicit sales can be seen to perform the same functions. As already noted, the UK was anxious to retain good relations with the Arab states which, with the exceptions of Libya and Syria, supported Iraq.

British officials and Ministers tried to make the case that the issue of diversion was not raised with Jordan because of the importance of the country to the UK.[111] Only the Ordnance Technologies case was raised with Jordan, but this was after the invasion of Kuwait by Iraq.[112] Some officials went so far as to suggest that enforcing British export controls would jeopardize the Jordanian State: 'The dangers of instability in Jordan can hardly be overestimated. We need to think very carefully before we say or do things which are going to cause unnecessary problems.'[113]

However, an official and a former Minister of State at the FCO both asserted that the relationship with Jordan was so close that the issue of diversion could have been raised by the UK without harm.[114] The Assistant Under-Secretary supervising the MED and the NENAD concurred with that view: diversion was not systematically tackled, he argued, because of the 'burden' it would entail upon officials.[115] But, bureaucratic resources will always be allocated to political priorities and he went on to suggest a more convincing reason why the Government failed to prevent the diversion of arms from Jordan to Iraq.

> At this time . . . we were actually mounting a very large Defence Sales campaign in Jordan. We were trying to sell them things rather than stop them getting things, and this was something that was being carried on at the highest level.[116]

Similarly, the content and tone of a record of a meeting between the British Ambassador and a Saudi prince, annoyed at Allivane's reneging on an arms contract destined for Iraq, suggests a similar concern for maintaining defence sales.

> Her Majesty's Ambassador had been briefed that it was not Her Majesty's Government's responsibility to question the Saudis' intention of using the firm instead of normal channels.[117]

It might be argued, therefore, that diversion to Iraq and Iran was ignored because of the increased determination with which the British

Government pursued arms sales in the 1980s. John Reed, the Editor of *Defence Industries' Digest* asserted that:

> From 1985 onwards, the entire face of defence exporting in the UK changed. In 1985 ... the Prime Minister had been directly involved in negotiating a sale of communications equipment to the United States Army and came away without an order, and ... this was something of a watershed. She realised the strength of the French companies in the market. A French company had beaten her and ... that was the moment at which the rules began to change.[118]

In short, the British Government allowed Jordan and Saudi Arabia to divert arms to Iraq both because of a general effort to increase sales, and a related desire not to jeopardize important defence contracts, and all that they might bring, with Arab states.

Illicit sales also served much the same purposes as the officially sanctioned trade. First, they kept open the markets for British defence goods in both Iran and Iraq. As a former Head of the DSO remarked: 'If over a period of time we adopt a policy of not supplying a certain country with defence equipment, we risk losing the market for a considerable period.'[119]

The Minister for Trade's concern in 1984 about restricting British defence exports might explain the UK's willingness to countenance grey market sales:

> We saw other countries supplying them without going through any of this rigmarole at all. I think there was a genuine feeling that here were very long-term risks that we would lose a great deal of *civilian* exports permanently when hostilities came to an end.[120] (emphasis added)

Second, there was an immediate cost to the UK in not supplying Iran and Iraq with arms, especially when many other countries were gaining shares of the lucrative market that was the Iran–Iraq War. As the then Foreign Secretary put it:

> the only questions arising were not 'Shall the Iraqis or the Iranians get this or that?' It (was): 'Shall we or shall we not stop British factories and workers from having the opportunity to supply?'[121]

British grey market sales to Iraq, given the support for the latter by both Britain's superpower ally and its Arab friends, are perhaps less in need of explanation than illicit sales to Iran. Already noted was Iran's importance to the UK in terms of trade before 1979 as well as, in the 1980s, its continuing *potential* significance as a market for British goods. On the other side of the equation was the vehement opposition to the supply of defence equipment to Iran on the part of the US and most Arab states. Illicit sales might be said to have bridged this conflict in British interests. Moreover, Britain's seeking of influence with Iran cannot be ruled out. Indeed, Aliboni saw all covert arms deals with Iran as: 'the expression of a widespread feeling, especially in the United States, that a working relation must be restored with Iran because of its unavoidable strategic importance'.[122] Even Saudi Arabia was reported in the press as attempting to supply arms to Iran.[123]

A further reason for allowing British companies to participate in illicit arms sales with Iran and Iraq was that the trade provided an important point of access for the intelligence agencies, especially in Iran after the 1979 Revolution. Such access would also have been an important tool in the management of UK relations with the US since the latter had lost its intelligence on Iran and, according to Paul Henderson, had scant resources in Iraq.[124] Henderson's assertion is supported by the SIS's sharing of his intelligence with the CIA from January 1989.[125] It is therefore possible that trade, especially illicit trade with both countries, represented an important medium of intelligence and this, in turn, represented a means of seeking influence with the US, given the historical role of intelligence in maintaining close links between the UK and the US. Of course, this is not to say that such intelligence was not useful in its own right in the quest for influence and trade. It might explain, however, why the UK did continue to supply Iran with defence goods in spite of US displeasure. A *quid pro quo* – *some* military trade for intelligence on Iran – might well have been in operation between the UK and the US.

Alternatively, the maintenance of Britain's links with Iran through the medium of illicit arms sales can be seen as either: a deliberate attempt, as Kenneth Timmerman argued, 'to step into a power vacuum created by a weakened US strategic position in the region';[126] or simply as a reflection of differing interests between the two allies. Certainly, Western Europe and Japan were less willing than the US to isolate Iran.

## British arms supply and the outcome of the Iran–Iraq War

In attempting to delineate any efforts to control the outcome of the war by Britain through its arms supply, it is necessary to sketch out the total supplies to Iran and Iraq by all countries.

| Iran | | Iraq | |
|---|---|---|---|
| Soviet Union | 40 | Soviet Union | 7,200 |
| Czechoslovakia, Poland, | | Czechoslovakia, Poland, | |
| Rumania | 50 | Rumania | 1,290 |
| China | 300 | China | 1,500 |
| US | 2,400 | US | 0 |
| France | 1,200 | France | 3,800 |
| Italy | 320 | Italy | 410 |
| Germany | 210 | Germany | 140 |
| UK | 575 | UK | 280 |
| Others | 550 | Others | 3,000 |
| | | | |
| Total | 5,365 | Total | 17,620 |

Current $ millions

Source: ACDA, *World Military Expenditures and Arms Transfers*, 1985, Washington GPO.

Table 6: Iranian and Iraqi arms transfers by major supplier, 1979–83

Tables 6, 7 and 8 give an indication of the paltriness of the British contribution to Iranian and Iraqi arms supplies. If the *type* of weaponry supplied by, *inter alia*, the Soviet Union, France and China is considered, one gets a sense of the insignificance of what was supplied by the UK, officially and via the grey market. China supplied, either directly or through North Korea, tanks and artillery; and the Soviet Union, like China, often using its allies to mask sales, supplied Iran with tanks and surface-to-air missiles. Iraq received tanks from China, and fighter aircraft, missiles and tanks from the Soviet Union.

Iraq was able to import arms with much greater latitude than Iran, spending between 1984 and 1988 nearly three times as much as Iran on military goods.[127] Iraq, during the course of the war, was supplied with $8 billion and $3 billion of Soviet and French arms, respectively, and $579 million of American defence exports.[128] Iran was increasingly forced, because of 'Operation Staunch' and its general isolation from the major weapons suppliers, to look to the black market, principally for spares for its mostly American arms inventory. Cordesman argued that:

> partly due to US influence, the total flow of Western arms exports to Iran dropped from $2.325 billion in 1983 and $2.9 billion in 1984 to $2.158 billion in 1985 and $860 million in the first half of 1986.[129]

Clearly, British arms could, at best – or at worst – only make a small contribution to the progress of the war when at least fifty countries were supplying either or both sides. Thus, if the point of British arms exports was either to prolong or to prevent victory by one side, the UK would have had to have acted in concert with other countries. First, then, British and Western interests in the course and outcome of the Iran–Iraq War need to be established.

## British interests in the Iran–Iraq War

After the 1979 Revolution, Iran was seen as the principal threat to the Gulf, especially since it sought to export its revolution using terrorism and subversion via its links with Shi'ite and radical groups. Shi'ites are the majority in both Iraq and Bahrain; they comprise 30 per cent of

| Iran | | Iraq | |
|---|---|---|---|
| Soviet Union | 5 | Soviet Union | 15,400 |
| Czechoslovakia, Poland, Rumania | 890 | Czechoslovakia, Poland, Rumania | 2,075 |
| China | 2,500 | China | 2,800 |
| US | 10 | US | 0 |
| France | 100 | France | 3,100 |
| Italy | 200 | Italy | 370 |
| Germany | 10 | Germany | 675 |
| UK | 100 | UK | 30 |
| Others | 6,705 | Others | 5,200 |
| Total | 10,520 | Total | 29,650 |

Current $ millions

Source: Anthony H. Cordesman, *After the Storm: The Changing Military Balance in the Middle East*, London: Mansell Publishing Ltd, 1993, p. 44.

Table 7: Iranian and Iraqi arms transfers by major supplier, 1984–88

| Country | Iran | | Iraq | |
|---|---|---|---|---|
| | Weapons | Other Support | Weapons | Other Support |
| *Supporting both parties* | | | | |
| Austria | x | | x | |
| Belgium | x | | x | |
| Brazil | x | x | x | x |
| Bulgaria | x | x | x | x |
| Chile | x | | x | |
| China | x | x | x | x |
| Czechoslovakia | x | x | x | x |
| Ethiopia | | x | x | |
| FR Germany | x | x | x | x |
| France | x | | x | |
| German DR | x | x | x | x |
| Greece | x | x | x | x |
| Hungary | x | x | x | x |
| Italy | x | x | x | x |
| Korea, North | x | | x | |
| Netherlands | x | | x | |
| Pakistan | x | x | x | x |
| Poland | x | x | x | x |
| Portugal | x | | x | |
| Saudi Arabia | | x | | x |
| South Africa | x | | x | |
| Spain | x | | x | x |
| Sweden | x | x | x | |
| Switzerland | x | | x | |
| UK | x | x | x | x |
| USA | x | x | x | x |
| USSR | x | x | x | x |
| Yugoslavia | x | x | x | x |
| | | | | |
| *Supporting Iran only* | | | | |
| Algeria | x | x | | |
| Argentina | x | x | (x) | |
| Canada | x | x | | |
| Denmark | | x | | |
| Finland | x | x | | |
| Israel | x | x | (x) | |
| Kenya | | x | | |
| Korea, South | x | x | | |
| Libya | x | x | | |
| Mexico | x | x | | |
| Singapore | | x | | |

| Country | Iran | | Iraq | |
| --- | --- | --- | --- | --- |
| | Weapons | Other Support | Weapons | Other Support |
| Syria | X | X | | |
| Taiwan | X | X | | |
| Turkey | X | X | | |
| Vietnam | X | | | |
| Yemen, South | | X | | |
| *Supporting Iraq only* | | | | |
| Egypt | | | X | X |
| Jordan | | | X | X |
| Kuwait | | | | X |
| Morocco | | | | X |
| Philippines | | | X | X |
| Sudan | | | | X |
| Tunisia | | | | X |
| United Arab Emirates | | | | X |
| Yemen, North | | | | X |

The term 'weapons' includes major weapons, small arms, ammunition or explosives.

'Other support' includes military transport vehicles (jeeps, trucks, lorries), spare parts, training, military advisers, logistic support or financial support.

Source: *SIPRI Yearbook 1987: World Armaments and Disarmament*, Oxford, Oxford University Press, 1987.

Table 8: Arms supply and other support to Iran and Iraq, 1980–86

Kuwait's population, and are a substantial minority in Saudi Arabia's oil-producing eastern province. The UAE has only a small Shi'ite population, but it is vulnerable to Iranian influence because of the volume of trade with Iran and its significant expatriate community of Iranians. Iran, by its actions and its example, in a region marked by oligarchy, disparities in wealth and a religion which does not distinguish between faith and government, threatened Western and Arab interests in the status quo.

At the same time, because of its closeness to the Soviet Union and its quest for Arab hegemony, Iraq was also perceived to be antithetical to Western interests in the Middle East. King Khaled is reported to have quoted a verse of Arabic poetry when told of Iraq's attack on Iran: 'Perhaps the snakes will die from the poisonous stings of the scorpions.'[130] The war, as Kissinger said, was the first in which America hoped both sides would lose.[131] However, Western countries frequently claimed to wish to see an early end to the conflict and 'Operation Staunch', George Shultz, the American Secretary of State, asserted, reflected the US policy of trying to stop the war.[132] British Ministers equally asserted that the UK's interest was in ending the conflict, and, indeed, this was the apparent premise of the Guidelines.

It is at this point that the ascription of interests and the interpretation of behaviour becomes very uncertain. The international community's response to Iraq's attack on Iran was somewhat muted. Some six days after the outbreak of war, the UN Security Council passed Resolution 479 which called for an end to the use of force and a peaceful settlement of the dispute. However, it did not make any reference to the fact of the Iraqi invasion or call for a withdrawal of forces to international borders. Moreover, it was another two years before the Security Council met to discuss the war again. Thus, on the one hand, there was apparent inaction but, on the other, account must be taken of the inability, until the late 1980s, of the superpowers to cooperate and Iran's intransigence.

Following less than a year after the invasion of Afghanistan, the outbreak of the war, argued Charles Tripp, raised Western concerns about Soviet intervention in the Gulf.[133] However, two months after Iraq's invasion of Iran, the British Foreign Secretary told the House of Commons that he thought the Soviet Union 'probably want to stop the conflict' and that they found it somewhat embarrassing.[134] While the risk of Soviet exploitation of the conflict remained, the West had two principal interlocking concerns: the security of oil and the containment of the conflict. As the war progressed and spread, at Iraq's instigation in 1984, to the Gulf waters, it became obvious that the conflict was having 'little sustained effect upon the flows of oil from Gulf producers'.[135] Furthermore, even though the threat of closure of the Strait of Hormuz, the narrow waterway through which Gulf shipping must enter, was projected to produce an immediate shortfall of world oil supplies of some 7mbd, provided the market did not over-react, the effect would be short-term. In any case, the threat was Iran's, and Iran had more to lose than to gain in closing the Strait.

Although, from 1984, the so-called 'Tanker War' did represent something of a hazard for shipping, it should be noted that, as far as the states involved were concerned, it was a tolerable danger, given that oil continued to flow from the Gulf and trade was uninterrupted. In all, Britain suffered only twelve attacks on her shipping, seven by

Iran and five by Iraq, with ten of these attacks occurring during Armilla patrol escorts.

In fact, the war had a beneficial effect in terms of oil prices. Because of excess capacity, the conflict acted as a prop to world prices. Having toured the Gulf states in 1987, the House of Commons Foreign Affairs Committee reported that:

> We found that the oil states were going through difficult economic patches as a result of the bottom falling out of the oil prices, a course which was likely to go further if the war was to come to an end and both sides were going to drop the price of oil in order to re-gear their societies.[136]

However, the concern for the price of oil was probably balanced, in the West, by the lure, at the war's close, of the inevitable reconstruction projects in both countries.

To what extent the West feared the spread of the Iran–Iraq War remains unclear. For example, in her memoirs, Lady Thatcher referred at the outbreak of the conflict to the 'potentially dangerous political and economic implications for Western interests' and said that 'these issues were to dominate British foreign policy in the years ahead'.[137] Thereafter she hardly mentions the Iran–Iraq War, except to note the arms sales to the Gulf arising from their sense of insecurity.[138] Thus, when the politicians spoke of the 'containment' of the Gulf War, it is perhaps better to see it as a code for the containment of Iran and Iraq: that neither country should conquer the other, and that Iran and its revolutionary message, in particular, should be confined to its borders. The Foreign Office was reluctant to spell out its desired outcome of the war. Pressed by the Foreign Affairs committee, the FCO official eventually said: 'An outright victory of a dominant kind by one party or the other would have a destabilising effect on the region.'[139] Another civil servant in the FCO added later that the UK was 'possibly more worried about Iran winning'.[140]

It is this ambivalence on the part of the UK and most other countries that has led to charges that, as O'Ballance put it:

> many nations have provided either Iran or Iraq with weapons, ammunition, military equipment, advisers, technicians, training, back-up facilities and services, some impartially to both countries, to enable both combatants to continue their military struggle. The unspoken proviso has been that a rough parity of arms should be

maintained between them, so that neither had sufficient means to defeat the other decisively.[141]

Thus while, on the one hand, the Foreign Affairs Committee spoke of the Iran–Iraq War as 'threaten(ing) the stability of the whole Gulf region',[142] other commentators saw the Arab states' security as relying upon 'the exhaustion option';[143] and the West's interests as lying in the war creating 'the conditions for Iraq's and Iran's future reliance on the major powers'.[144]

As far as Britain was concerned, Alan Clark, the former MinDP, maintained that 'Britain's real interest lay in that Iran and Iraq should bleed themselves almost to death but never actually crash to the ground.' Neither country could be allowed to 'crash to the ground' because the hegemony of either would be anathema to Western interests and, in the event of an Iranian defeat was the fear that 'the Russians would have probably taken a chunk of it, come over the Caucasus'.[145]


### The British capability to affect the Iran–Iraq War

Even if Western and Arab countries had specific interests in the prolongation or the eventual outcome of the Iran–Iraq War, these interests are meaningless without the wherewithal to effect the desired end. On this point, the nature of the conflict is important.[146] It was the limited nature of Iraq's attack on Iran, together with the lack of initiative exercised by the Army, that quickly led to stalemate in the war. From 1982, the Iranians took the military initiative, and regained most of their territory. On the defensive, the Iraqi Army found a renewed morale and were able to withstand Iranian offensives. Until late 1985 the ground war was essentially a war of attrition in spite of Iraq's escalations in beginning both the 'Tanker War' and the 'War of the Cities'. It was only in 1986 and 1987 that Iran began seriously to threaten Iraq with successful offensives in Fao and Kurdistan and with a large but ultimately unsuccessful offensive towards Basra. However, between April 1987 and March 1988, the anticipated 'final offensive' by Iran did not materialize. The war ended when Iraq restructured and retrained its forces, regaining its own territory and pushing into Iran in the Spring of 1988. For its part, Iran could no longer muster the manpower to meet the Iraqi offensives as popular morale had slumped. Checkmated on the ground, and at sea by the Americans, Iran was forced to accept UNSCR 598.

Summing up the conflict, Sir John Moberly commented:

It has been fashionable to say that Iraq cannot win the war but could lose and that Iran cannot lose the war but could win, though precisely what the terms win or lose might mean could be the subject of much discussion.[147]

Any discussion of the use of arms supplies to effect a given outcome in a war must be held against the backdrop of the essential uncertainty of warfare – Clausewitz's notion of friction. What must also be borne in mind is that, on the one hand, the war was, for the most part, attritional and, on the other, that there existed between Iran and Iraq 'a delicate balance of incompetence'.[148] SIPRI have pointed out the significance of arms supply during conflict:

An imbalance in arms supplies to two parties with similar technical resources could lead to the speedy conclusion of a war, while a correction of imbalance could prolong it.[149]

However, as Olson pointed out:

Although Iran used largely US weapons and Iraq used largely Soviet equipment, neither belligerent employed them the way the US or the Soviets would have. Neither belligerent made effective use of either air or ground forces; this is particularly true for Iraq, which enjoyed . . . virtually absolute superiority in every area of combat.[150]

For Iran, being forced on to the black and grey markets for arms, ammunition and spare parts for its largely American weapons meant that it could 'assemble only part of the balanced mix of supplies, spares and weapons it need(ed) to support a major offensive'.[151] Moreover, it suffered from being unable to procure fighter aircraft and surface-to-air missiles. However, Iraq's superiority in arms could not compensate for poor training, a dearth of initiative and lack of numbers. On a visit to Jordan in 1986, the Minister of State at the FCO was told of the Jordanian Commander-in-Chief's puzzlement about Iraq's failure to exploit its superiority in equipment.[152] It was, as Cordesman pointed out, only when Iraq learned the importance of infantry tactics late in the war, that it was able to gain a decisive military superiority over Iran.[153] Thus, if decision-makers appreciated the above uncertainties, then their intentions with regard to effecting any given outcome in supplying either or both sides must be a matter of doubt. Their calculations would have been further compromised by the existence of so many suppliers.

Participants in the illicit trade with Iran and Iraq have argued that, in the words of Astra's Chairman, Gerald James:

> The cartel structures could not have existed and operated without the full knowledge, direction and co-operation of the intelligence and security services: MI5, MI6, GCHQ, JIC and their counterparts in other Western countries, particularly the Defence Intelligence Agency and the CIA.

Gerald James went on to allege that the direction of the illicit arms sales 'was determined by the intelligence services aided by certain Ministers, industrialists and bankers'.[154] By its very nature, the evidence underpinning such assertions is circumstantial. For example, Astra was encouraged to buy PRB by Sir John Cuckney, Chairman of IMS until 1985 and Chairman of Astra's main shareholder, the 3i investment bank, and by a non-executive director of Astra, Stephan Kock. Both men had close links with the intelligence agencies as well as with business interests and the Government. Kock was also said to be involved in the Sweden-based cartel which supplied both Iran and Iraq.[155] After the purchase of BMARC, Jonathan Aitken MP pressed James to make him a member of Astra's board of directors. Both Gerald James and Paul Henderson argued that Roy Ricks, the British partner in the company, MEED International, which headed up Iraq's procurement network in the UK, was an intelligence agent.[156] Certainly Ricks, together with Paul Henderson, was identified by intelligence officers in September 1988 as a possible agent with access to Iraqi-owned companies in the UK.[157]

If the British intelligence agencies were involved in the direction of supplies to both Iran and Iraq, then their overseas counterparts probably knew of the activity. Ed Juchniewicz, an American official involved in 'Operation Staunch', recalled that the CIA identified the channels used by Western states to supply Iran with arms.[158] George Shultz, the US Secretary of State, asserted that, when American arms sales to Iran became public, Admiral Poindexter, the National Security Council adviser, said that the US's European allies 'all traded with Iran anyway, so they couldn't complain about what we were doing'.[159] However, while this establishes knowledge, the tone and content of the remark does not seem to indicate approval, although one cannot, by definition, rule out the possibility of back channels operating both within and between the US and the UK.

The evidence, however, seems to indicate that there was no international conspiracy, tacit or otherwise, to control the progress and outcome of the Iran–Iraq War through arms supply. First, the war was an embarrassment to the Soviet Union: it punished Iraq for its invasion in 1980 by withholding arms supplies, and only resumed sales for fear of Iraqi capitulation. Second, the US, from 1983, saw *ending* the war as soon as possible as the only way it could prevent an extension of Soviet influence in the Gulf and, more importantly, avoid an Iraqi collapse. Third, as previously noted, the effect of the supply of arms, in the specific circumstances of the Iran–Iraq War, could not be predicted, and this ought to have been clear to all suppliers at a fairly early stage of the war.

Alone, the UK was not going to affect the progress or outcome of the war. While, as Harkavy argued, attrition warfare increases the importance of arms resupply,[160] such resupply, given the existence of other, larger arms suppliers, is more likely to have served political interests because it represented an important signal of support. What was officially supplied, Chieftain spare parts for example, could only have been locally and temporarily important. Because of the quantities involved, British exports were not going to tip the balance in favour of either participant, given that the war was one of attrition.

While grey market sales of explosives, ammunition and arms added to the British contribution, here again the significance of the supplies would have been limited because, as Alan Clark put it, 'The trouble with under the counter supplies is that they cannot be as prolific as straight stuff.'[161] Since the UK was trying to maintain relationships with all sides, it had to limit illicit supplies to types and quantities of weaponry that could be passed off as failures in control mechanisms because, as Stanley and Pearton put it, 'arms sales have the unruly habit of shattering both secrecy and ambiguity'.[162]

While a note of caution should be added here in that the true scale of British grey market sales to Iran and Iraq is not fully known, other evidence points to the UK using arms supply as simply a tool of influence with the belligerents, rather than as a means of affecting the progress of the war, However, the evidence remains ambiguous.

## A 'tilt' to Iraq?
Both Alan Clark and Gerald James, the former Chairman of Astra, assert that Britain, from 1987, concerned at Iran's preponderance,

began to support Iraq much more.[163] Certainly, in January 1987, the US became even more alarmed by Iran; and British policy is responsive to American goals. As the then US Secretary of State recalled:

> Iran had launched a major offensive, driving into Iraq just north of Baghdad and south of Basra, near the Persian Gulf. Iran was mining the Gulf, menacing commercial shipping, and threatening to close the Gulf and strangle the states of the Arabian Peninsula, let alone Iraq. Iran sought to position itself to dominate the entire region.[164]

In fact, it was in July that Iran began to mine the main shipping channels in the northern Gulf and in August that it mined the approaches to the Straits of Hormuz, although it had increased its attacks upon shipping throughout the year and had attacked Kuwait with Silk Worm missiles.

The British do seem to have shared the alarm expressed by the Americans at this time. In his *Diaries,* Alan Clark recalled that for almost the last two years of the war, 'the FCO section of Cabinet minutes was a long moan about how the Iraqi Army was on its last legs, and the Iranians were going to break through'.[165] It is also apparent that the British, in common with other European powers, feared that the US might seek a direct confrontation with Iran.[166]

Thus it was that the American request for British mine-sweeping assistance was initially turned down. This, argued David Sanders, represented 'the first time under Thatcher, the British Government openly demurred'.[167] However, the evidence here suggests that the UK had been demurring in its policy towards Iran since the outbreak of the war. Fearing the US military course, but at the same time sharing a sense that the escalation in the war threatened Iraq's survival, and possibly that of other Arab states, as well as Gulf shipping, the UK seems to have struck out on a more diplomatic course. It was instrumental in securing the adoption of UNSCR 598. This was the first Security Council Resolution which tackled Iran's insistence upon establishing and acknowledging the aggressor in the conflict. Conversely, as a mandatory resolution under Articles 39 and 40, the threat of an arms embargo – more damaging to Iran than Iraq – was implicit. As Sir Anthony Parsons put it: 'The only place really we can demonstrate our policy in regard to the war is in the forum of the UN . . . and in the long-term, it will serve our interests too.'[168]

By this point, the UK had joined the US, the Soviet Union and France in having so much invested in Iraq that an end to the war became more urgent. Iraq's indebtedness to each of them had given the four countries an additional stake in bringing the conflict to a close. Iraq's arms suppliers were reported to have threatened to suspend exports in 1984 and in 1987 because of unpaid debts.[169] By 1987, France had acquired some $5 billion in unpaid debt[170] and the US almost $2 billion.[171] Iraq's total debt to the UK was $1.3 billion and Iraq was in arrears with its debt payments to the UK. Allied to the concern about Iraq's ability to meet its debt obligations was the trade which these states had developed from the late 1970s. The US was selling $1 billion of wheat alone to Iraq each year.[172] While the UK's interest was less significant, Iraq had nonetheless become an important market; by 1986, the UK was Iraq's second largest non-military supplier.[173] As a senior DTI official put it: 'from looking at intelligence about the general size of (the) UK plc stake in the market in a whole variety of ways . . . it was not going to be in Her Majesty's Government's interests to see Iraq go down the tubes'.[174]

What is significant, for the purposes of this analysis, is that the UK appears to have sought to effect the desired outcome, the ending of the war, through diplomatic channels, rather than through arms supply. Although spares for the Iranian Navy had been curtailed in September 1987, following the attack on the *Gentle Breeze*, other defence equipment continued to be supplied to Iran under the Guidelines.

However, the evidence remains unclear. The fact that, in 1988, defence exports to Iran were worth only £1 million might be representative of the UK's desire to end the Iran–Iraq War by putting pressure on Iran, or it might be the result of Iranian trade sanctions against the UK instituted in 1987. The sale to Iraq of machine tools for the manufacture of artillery shells in 1987 and 1988 might constitute evidence of a 'tilt' in Iraq's favour, but machine tools for munitions production had also been sold to Iran. A 'tilt' is far from obvious in Anglo–Iraqi discussions, from May to September 1987, in which the British made clear that the use of additional credit was dependent on arrears being kept within certain limits.[175]

## Summary

The bulk of the evidence which has thus far emerged points to Britain being motivated by overlapping reasons of influence and trade in its arms export policy towards Iran and Iraq. Arms supply was aimed at the belligerents, rather than at the war. For example, the so-called tilt towards Iraq in 1981 and 1982 seems to have been, as the then Under-

Secretary for the Middle East, Sir Stephen Egerton, put it, 'for commercial reasons'.[176] Certainly, the minute of the Defence and Overseas Policy Committee of the Cabinet meeting which recorded the desire to exploit Iraq's potential as a market for British defence goods supports that thesis.

Sir Stephen went on to say that the British Government did not share 'the overwhelming Arab preoccupation' that Iraq might be defeated.[177] Thus the UK's continued arms supply to Iran in the face of US and Arab opposition can be interpreted as being premised upon a belief that what was supplied was not going to be ultimately decisive and thus the UK was free to pursue the tangible interests of maintaining relationships and pursuing trade. As the former Foreign Office Minister put it, on the one hand the British Government 'had no means of actually having much influence on events',[178] and, on the other, as the MinDP wrote in 1984, the UK did 'not wish to totally sever the residual defence links' which it had with Iran because it was 'important that relations of a more general sort (were) maintained'.[179] The UK was not alone, even among the Arab states,[180] in wishing to retain a connection to Iran. Therefore, while British arms might not have been able to tip the balance in the conflict, they remained an important instrument of foreign policy.

While the UK's arms export policy remains open to interpretation, the most plausible motives behind the totality of British arms export policy were the overlapping desires for influence and trade: the war lasted so long because, as O'Ballance pointed out, 'both combatants had the means to purchase weaponry, as both had huge reserves of oil and natural gas to sell, or use as barter'.[181]

# 6

# Iraqi proliferation

## The export of ambiguous goods

A study of the motives behind Britain's supply of dual-use goods to Iraq reveals the same forces at work as in the export of overtly military equipment. If, as the academic consensus has it, the UK, as a second-tier state, is forced to sell arms to mop up spare capacity in the defence manufacturing base, then one would expect to find less pressure to export dual-use goods for projects known or suspected to be, military, and especially non-conventional, in nature, given the sheer range of companies involved and the scope of the market available to them. However, it was the more general concerns for trade, and the relationship upon which that trade depended, which drove the sale of dual-use goods. It is in the UK's activity, or inactivity, regarding Iraq's procurement network that the strength of these motives can be seen. While the potential costs of supplying Iran and Iraq with conventional weapons were political and economic, there were strategic implications in allowing Iraq to procure dual-use technology.

Given the nature of dual-use equipment, transfers of arms-making technology can represent a conscious decision by a government or can be inadvertent. In the former case, they are part of Klare's definition of 'grey market' sales.[1] Either way, such sales not only diminish the technological advantage of the supplier but also, as Klare pointed out, increase the competitiveness of the arms market by creating more suppliers.[2] Technology transfer, in contributing to the indigenous arms manufacturing of recipients, makes the strategic environment more complicated and less malleable to the industrialized powers. As J. Fred Bucy noted:

the release of know-how is an irreversible decision, Once released, it can neither be taken back nor controlled. The receiver of know-how gains a competence which serves as a base for many subsequent gains.[3]

More importantly, the sale of dual-use goods to states procuring them for nuclear, biological, chemical or ballistic missile programmes has profound global strategic implications.

Sales of dual-use items represent the loophole in the international control regimes set up to counter proliferation in weapons of mass destruction and the means of delivering them. However, control of such goods is complicated by a number of factors. First, if the potential recipient spreads its procurement net widely enough, it can be difficult for potential suppliers to accurately perceive true intentions. Second, there is, by definition, an intrinsic uncertainty about the end use of such exports, which even intelligence might not be able to counter. Finally, since the rewards of selling are immediate, and the costs are both potential and long-term, given the first two uncertainties, the disposition of governments seems to be to export where no international control regime exists.

The demand for technology transfer, in all its forms, 'grows stronger by the year'.[4] Although recipients remain dependent upon imported technology, this dependence is, as Ross argued, 'less burdensome than depending on imported weapons'.[5] The reasons why states seek to develop indigenous weapons manufacturing capability are at once complex and obvious involving issues of national independence and local economics. In Iraq's case, the initial spur to the development of an industry, in both conventional and non-conventional weapons, seems to have been the Soviet Union's manipulation of ammunition supplies in 1975. This led to Iraq's inability to contain the Iranian-backed Kurdish insurrection and, from there, to its conceding half the Shatt al Arab waterway to Iran. However, the construction of an arms industry was also part and parcel of the more general development of a manufacturing base. The war with Iran naturally increased the impetus to develop an indigenous defence capability especially when, in 1984 and again in 1987, Iraq's principal suppliers threatened to suspend arms transfers until outstanding debts were paid.

Increasingly, throughout the 1980s, the UK contributed to that programme both in conventional and non-conventional weapons. In making such a contribution the UK was not alone: most Western countries became drawn into the Iraqi procurement network. The International Atomic Energy Agency (IAEA) reported in 1991 that:

Among procurement strategies employed were: the use of other Iraqi establishments as buyers and contractors; the placing of orders for equipment directly with manufacturers or indirectly through foreign intermediaries; and the use of indigenous capabilities to complete the manufacture of some items.[6]

Thus, in explaining British participation in Iraq's procurement strategies, the question of government knowledge about Iraqi intentions lies at the heart of the matter.

A complete assessment of what Britain supplied to Iraq's weapons programmes is not yet possible, nor is it necessary for the purposes of uncovering motive. What is necessary is to show whether or not goods were *knowingly* supplied. The principal examples chosen for study in the succeeding chapters are exports of machine tools and exports of parts of Project Babylon, the supergun. Machine tools are chosen because they represented dual-use technology and contributed both to conventional arms manufacture and to Iraq's nuclear weapons programme. The IAEA reported that thirteen of the forty-seven key machine tools in Iraq's nuclear weapons programmes were supplied by the British company, Matrix Churchill.[7] The company was found to have made components for a centrifuge used in uranium enrichment for the Iraqi nuclear weapons programme, Project K1000,[8] as well as contributing machine tools to the programme which modified *Scud* missiles. The supergun case is chosen in order to illustrate the contention here that, where the national security of Britain or its allies or friends was at stake, the Government acted to prevent exports. It is where there was ambiguity, or where the risks of supply were in the long term, that other considerations came into play. It should be noted, however, that the only ambiguity about the supergun tubes lay in the companies not applying for ELAs for *gun barrels*; but this ambiguity was enough to allow the export to proceed.

Fundamental to an explanation of both the supergun and machine tools affairs is the issue of government knowledge. In order to establish responsibility, and therefore policy, one needs to ascertain what was known, by whom and when. The Government has tried to make the case, in both affairs, that because knowledge was fragmented, so too was responsibility. In short, the export of both machine tools and parts of the supergun represented a failure in processes rather than an expression of policy. This was not the case. However, this is not to suggest that it was the Government's policy to assist Iraq's programmes in weapons of mass destruction. As this book has tried to establish, in many cases export licensing decisions represent a contest between competing economic, political and strategic considerations. In the trade-off between these different concerns, or different policies, the contention here is that, where no international control regime exists, and equipment or end use is ambiguous, both economic policy, the impetus to export, and foreign policy, the desire to nurture relationships, will be stronger than long-term strategic policy, the desire to prevent proliferation.

## 'UK Ltd' and Iraqi proliferation

The chosen examples should not obscure the fact that Britain's contribution to Iraq's weapons programmes was altogether wider than this narrow selection might suggest. The US Department of the Treasury listed some fifty-two companies worldwide that had knowingly or unknowingly contributed to Iraq's procurement network by being owned by, or controlled by, or acting on behalf of, the Iraqi Government. Of these companies thirty-three were based in the UK.[9] While the Treasury list was criticized by the UK Government on the basis of its inaccuracy and some eight British companies met DTI officials in an attempt to clear their names,[10] the UN passed to the Foreign Office the names of twenty British firms which had helped equip Iraq's programmes in weapons of mass destruction.[11] The scale of the UK's contribution to Iraq's weapons industry was noted in an MOD assessment from the late 1980s, the purpose of which was 'to draw the attention of the Minister to the way in which "UK Limited" is helping Iraq, often unwillingly, but sometimes not, to set up a major indigenous arms industry'.[12]

There were two unavoidable omissions from the MOD's assessment. Only after 1987 did the DTI refer industrial list goods for Iran and Iraq to the MOD; and items exported on temporary licences which were not returned provided 'another loophole' in the control of exports to Iraq.[13]

Thus, the chosen examples should neither be seen as *sui generis* nor as representing the most significant items supplied. Kenneth Timmerman argued that, apart from machine tools, it was advanced computers and the building of an entire semi-conductor manufacturing plant which comprised the most important British contributions to Iraq's NBC and ballistic missile programmes.[14] The MOD's study highlighted:

1. A major R&D facility.
2. A national electronics manufacturing complex.
3. Foundries designed to produce special steel for gun barrels and tank parts.
4. Machinery to make gun barrels and shells.
5. Specialist factories/production lines for parachutes or thermal batteries.[15]

Among the individual items pinpointed by the assessment were the technology transfer and supply of equipment for the manufacturing of moulds for steel ingots and the computer and software for checking the accuracy of precision moulds.[16] These items, as well as the thermal batteries production line, all have uses in the manufacture of missiles. British computer software was reported as being used in the injection moulding of plastics in Iraq's missile programme.[17] The MOD study also highlighted the supply of vacuum precision furnaces to Iraq on the grounds that they are 'capable of manufacturing aircraft engine components'.[18] However, they are also capable of making missile components and, on this basis, the US sought, unsuccessfully, the co-operation of the UK in preventing their export.[19]

The Government released a list of chemicals exported to Iraq to the Trade and Industry Committee. However, the list only covered the period from October 1988 to June 1990. Records regarding chemical exports were said to have been destroyed for the period to February 1988.[20] Thus a complete assessment of the UK's contribution to Iraq's chemicals weapons programmes is impossible to conduct. Dr Alastair Hay, a chemical pathologist, in a letter to the Trade and Industry Committee, noted that, 'The United Kingdom sold shipments of sodium cyanide and sodium sulphide to Iraq when it should not have done so.'[21] The chemicals were exported, legally, to Iraq in 1988, 1989 and 1990: they were brought under statutory control in July 1991 but have been on the Australia Group's Warning List since May 1989. In January 1988, some 48,000 kg of sodium sulphide was exported to Iraq, but to what establishment is unknown since the DTI no longer had the records.[22] Sodium sulphide has uses in various manufacturing processes, but is also an agent in mustard gas. Sodium cyanide again has various uses in civil industry, but is an agent for Tabun, used by Iraq in 1984,[23] as well as for hydrogen cyanide and cyanogen chloride. Three shipments of these chemicals were made after they had been added to the Australia Group list.

Dr Hay made no reference to the tri-ethanolamine shipped to Iraq in

1988. This was brought under UK statutory control in 1989 after it was placed on the Australia Group List. Dr Hay concluded his letter to the committee:

> it would seem that the Department of Trade and Industry will not be able to tell us whether Britain exported any of the fifty chemicals on the Australia Group lists to Iraq between 1980 and 1988. It would have been reassuring to know that Britain did not contribute in any way to the development of Iraq's chemical warfare programme. It now looks as though we will never know what was sent.[24]

Moreover, one will never know what was diverted to Iraq from *inter alia* Egypt, Saudi Arabia and Jordan. Some months after hydrogen fluoride was placed on the Australia Group Warning List and brought under British statutory control, a shipment of 26 tons was permitted to Egypt. Yet, it was intelligence regarding the Egyptian purchase of 'substantial quantities of hydrogen fluoride on behalf of Iraq' in January 1986[25] which led to statutory control and its addition to the Australia Group List.[26] However, the 1986 shipment went with a strong warning to the Egyptians that the UK expected the chemical not to be diverted from the 'legitimate civil applications' of the Ministry of Military Production.[27]

Already noted is the wealth of evidence about the diversion of military goods by Iraq's allies. This diversion encompassed the dual-use technology used in WMD as well as conventional arms. In 1984, a DIS Loose Minute noted the 'considerable' evidence of the 'links between Egypt and Iraq in the chemical warfare field' as well as between Jordan and Iraq.[28] The effectiveness of the quota system, in force from 1987 for NBC equipment to Egypt, Jordan and Saudi Arabia because of fears of diversion to Iraq, remains unknown.

The British contribution to Iraq's weapons programmes also extended to the provision of training as well as equipment. A total of 430 Iraqis received training at MOD installations between 1981 and March 1990.[29] In addition, Iraqis studied at British universities. In 1990, thirty-six members of the Iraqi Armed Forces were admitted to the UK to study at various colleges.[30] The number of British scientists and engineers who assisted Iraq's procurement projects is not known, but a 1990 government assessment of trade with Iraq drew attention to the importance of invisible trade 'particularly consultancy'.[31]

Given the various British contributions to the development of Iraqi WMD, it is odd that the one case where exporters have been jailed should involve goods which were not, as the prosecution claimed, destined for an Iraqi nuclear weapons programme. In 1990, in a joint Federal Bureau of Investigation–Customs operation, electrical capacitors were seized on the grounds that they were nuclear triggers. Mrs Thatcher congratulated Customs upon their actions.[32] Dr John Hassard, a nuclear physicist who had seen the UN documents relating to the Western contribution to Iraq's weapons programmes, asserted that the capacitors were not used in the nuclear weapons project. Furthermore, he argued that the capacitors were 'two hundred times too slow for the nuclear weapon as described in (the UN) documents'.[33] Those prosecuted had their convictions quashed on appeal.

Three possible explanations for the 1990 seizure and subsequent prosecution suggest themselves: first, it is possible that what was meant was a warning to Iraq; second, it was meant to reassure the public that the Government was active against the procurement network; and third, an honest mistake was made.

## The Government's knowledge of Iraqi procurement

Before examining the specific case studies, it is useful to establish the general state of the Government's knowledge about Iraq's military procurement programmes. Sayigh argued that:

> the first indication of the existence of a significant indigenous defence production capability in Iraq came with the employment of chemical weapons against Iranian troops in 1984.[34]

However, Timmerman asserted that the first indication of Iraq's intentions in this respect came in 1976 when ICI informed the British Government that it had been approached by Iraq for 'sensitive' materials which had 'the potential for misuse'.[35] In 1984, in common with other Western powers, the UK responded to Iraq's use of chemical weapons by bringing eight precursor chemicals under statutory control. Further precursors were brought under control in successive years following the establishment of the Australia Group, and the further use of chemical weapons by Iraq.

By 1984, the UK would have been aware, through the reports of business people if nothing else, about the scale and range of Iraq's indigenous arms industry. The threat of interruption to Iraqi arms imports occasioned by debt in the mid-1980s added to the impetus for military manufacturing. Timmerman argued that between 1985 and

1989, Iraq spent $142 billion on military-related technology[36] – that is, about half of all spending on arms imports.

If the British Government missed these developments, either in the export licence applications received or in the reports of business people, diplomats and Military Attachés, then events in 1987 would have given a clear indication of the scale of Iraq's defence industry. First, the Baghdad Fair in November 1987 contained an exhibition of Iraq's manufacturing capabilities.[37] Second, and more importantly, in August 1987, Iraq tested a medium-range ballistic missile, the *al-Hussein*, a modified *Scud-B*, with a range of 650 km.[38] These were used against Iran in February 1988. This development seems to have been taken seriously enough by the British to amend the Export of Goods (Control) Order on 1 January 1988 to bring a widened range of missile and space-related products under control for Iraq.[39] Third, from the summer of 1987, the Sales Director with Matrix Churchill had been providing intelligence about Iraq's purchases of European machinery for use in arms manufacture.

From October 1987, the British Government were aware of companies used by the Iraqis in the UK because of the purchase of Matrix Churchill by an Iraqi holding company, TMG, on behalf of Technology Development Group (TDG) and Technology Engineering Group (TEG), both Iraqi front companies. The CIA was reported as having disseminated intelligence about Matrix Churchill to policy-makers in December 1987.[40] Intelligence about the Iraqi procurement network from Matrix Churchill continued into 1990 and included information about: BNL's role in funding Iraqi purchases; SRC's plans for an artillery fuse factory; and a bomb factory provided by a Chilean arms manufacturer, Carlos Cardoen.

The British Government's knowledge of the extent of Iraq's efforts in the UK is demonstrated by the contents of departmental correspondence about a visa application for Dr Saafa al-Habobi in the Spring of 1988. An official in the FCO wrote to his superior and to the Minister of State to say of Dr al-Habobi: 'His strategy in essence seems to be to use TDG and TEG and their dependent UK companies to supply machine tools and other equipment for large scale munitions manufacturing in Iraq.'[41] The fact that it was the Foreign Secretary who wrote to the Home Secretary to request that Dr al-Habobi be given a visa, in spite of 'the extent of his business involvement in the UK' and 'his position in Iraq', indicates that the knowledge of the procurement network was not confined to officials or junior Ministers. The Foreign Secretary argued that the UK should take the 'less certain method of foiling al-Habobi's plans' by 'close monitoring of his activities', 'careful scrutiny' of export licence applications and making 'equipment licensable if necessary'.[42]

It was because the latter two recommendations were not implemented that the MOD security staff, in conjunction with the DIS, set out to study the scale and nature of British assistance to the Iraqi arms industry. Parts of the MOD had become concerned that some militarily useful equipment was not licensable. This concern however 'was not reflected in the Department of Trade and Industry . . . and it was only if we had managed to enlist Ministerial support would we have (had) the power to be able to make a change'.[43]

The study was first circulated within the MOD on 26 June 1989. Other Departments knew of its existence at that time, although it was not circulated to them in 1989 because it was awaiting a DIS Annex. However, the Chair of the MODWG, a senior member of the DESS, argued that the paper was 'going to add very little to what was already known'.[44] From July 1989, the MODWG began marking suspect ELAs as 'R(AM)' or 'Refused Arms Manufacture', also alerting other Departments to the Iraqi procurement network.[45] But, again, as the Chair of the MODWG put it, in defending his inactivity in regard to securing the DIS Annex, 'Everybody (knew) all about this indigenous arms manufacturing intention.'[46]

In November 1988, senior FCO and SIS officials decided at a meeting that 'SIS should put Iraq at the top of its priorities': the fear was that Iraq was to acquire 'a poor man's nuke', a ballistic missile fitted with a chemical warhead.[47] But, in 1988, the Government learned that Iraq was attempting to procure components and technology from the UK for its nuclear and ballistic missile programmes. In December 1988 Dr al-Habobi was implicated in Iraqi attempts to procure equipment for the development of gas centrifuge technology for uranium enrichment. The FCO noted:

> This is a serious development which confirms our long-held suspicions that Iraq, although a party to the Treaty on the Non-Proliferation of nuclear weapons (NPT), has ambitions to develop a nuclear weapons capability.[48]

In 1988 and 1989, the British Government was gaining 'a growing volume of evidence'[49] on Iraq's ballistic missile and nuclear capabilities. Such was that evidence, that when the Prime Minister saw it, she requested that Departments take co-ordinated action against Iraq's procurement network.[50] As a result, the Working Group on Iraqi Procurement (WGIP) was set up on 16 May 1989. The WGIP was essentially the REU meeting just before its parent body. At its first meeting, the record noted that:

> Early briefing has made it clear that the network is largely sophisticated and well-financed. Efforts to frustrate its acquisition of nuclear related technology are likely to be long-running.[51]

If nothing else, it was Iraq's purchase of the Learfan factory in Northern Ireland in the Summer of 1989 which alerted the whole Government to Iraq's quest for long-range missile technology. The Learfan factory was desired, to quote the Foreign Office Minister, for its 'MTCR technology'[52] i.e., carbon fibre technology. However, intelligence reports throughout 1989, circulated widely within Whitehall, made clear Iraq's capabilities and intentions in procuring weapons of mass destruction and ballistic missiles. In September 1989, intelligence reports identified a large Iraqi missile project headed by Dr al-Habobi. The project was a missile other than the Condor, and British companies were identified as being involved.[53]

Also noted by the Government was an article in *The Times* in November 1989 on the development of the Condor II long-range missile. The press had begun reporting on Iraq's procurement network in 1988. This article was a highly detailed description of how Iraq, Argentina and Egypt, the partners in the project, used a network of European companies to bypass the MTCR. The article also gave details of BNL's financing of the project, as well as asserting that Iraq's capability was so far advanced that it was able to proceed alone with the development of the Condor II.[54] At the December 1989 meeting of the IDC, the record, copied to the Minister of State, reported 'the launch of the Iraqi space vehicle'.[55] The launch of the *Tamouz* was reported in the press in February 1990.[56] On 24 May 1990, the Overseas and Defence Committee of the Cabinet discussed TDG's bid to purchase the Armadale steel foundry in Scotland. The DTI's paper for the meeting expressed the view that Iraq wanted to produce specialized castings for its nuclear industry.[57]

Thus, from 1987, the British Government was well aware of Iraqi intentions to develop an across-the-board defence manufacturing capability. Decisions were made against the backdrop of this knowledge.

# 7

# The supergun

## Differing accounts

The supergun affair is, as an official put it, 'rather a messy story'.[1] A detailed consideration of the narrative is necessary, not least because three versions of events exist: the Government's original position; the case put on behalf of the British companies involved; and the account produced by Sir Richard Scott. The House of Commons Trade and Industry Committee also investigated the affair. The Scott Inquiry had access to evidence unavailable to Parliament, yet it leaves many unresolved issues. The various accounts are summarized in this chapter and the remaining puzzles discussed. Judged in its entirety, the evidence points to a political decision to allow exports of parts of the supergun in 1988 and 1989; either because of an absence of concern about what appeared then to be a conventional weapons project, or because the exports represented another window of intelligence onto the Iraqi network. However, the eventual seizure of the final consignment of gun barrels indicates the UK's concern to act where its, or its friends', national security is threatened.

On 10 April 1990, Customs officers at Teesport detained eight large steel tubes destined for Iraq. The tubes did not have export licences. Examination by experts from the MOD confirmed that the pipes were 'components of a large calibre armament, albeit of a scale outside anything previously experienced'.[2] They had been procured from Britain by Space Research Corporation, a Belgium-based company owned by Dr Gerald Bull.

Bull was an engineer of genius obsessed with launching a space rocket from a gun. He had worked on such a project in the early 1960s. Those who knew Bull argued that, in 1987, the Iraqis wanted his expertise in artillery: the supergun project was the price they were willing to pay for that expertise. Bull was murdered in March 1990 by an unknown assassin.

### The Government's case

The Government claimed that the companies concerned, Walter Somers and Sheffield Forgemasters, had 'used the number of a genuine petrochemical project known as PC2', orders for which had been promoted through the British Overseas Trade Board in consultation with the Embassy in Baghdad. Since, according to the Secretary of State for Trade and Industry, 'It is possible to export material for petrochemical works freely, and no export licence is required',[3] the Government's defence lay in the deception practised upon it by the companies and in the 'compartmentalism of information' within Whitehall.[4]

The Government did acknowledge some contacts with the companies but disputed the nature of them. In June 1988, according to the then Secretary of State for Trade and Industry, Walter Somers asked if licences were needed for the export of metal tubes to SRC in Belgium and, the following month, Sheffield Forgemasters asked if licences were required for the sale of tubes to Iraq for use in the polymerization of polyethylene. In August 1988, Walter Somers made another, written inquiry related to that contract. Forgemasters wrote to the DTI in August 1989 and the DTI responded in November.

The DTI argued that it 'had no knowledge that the goods were designed to form part of a gun' until April 1990.[5] The MOD admitted that two telephone conversations had occurred between Walter Somers and an MOD metallurgical expert, Bill Weir, in 1988.[6] The Government further asserted that it had no knowledge of the existence of Project Babylon until September 1989.

The Government asserted that it was a lack of information which led the DTI to tell the company that no export licence was required.[7] The DTI's evidence to the Trade and Industry Committee was that Sir Hal Miller MP, acting on behalf of Walter Somers, had telephoned merely to ask if export licences were needed for the metal tubes to be produced by the company who were sharing the Iraqi contracts with Forgemasters. Somers made the Baby Babylon, the prototype, tubes, Forgemasters, Babylon. Sir Hal did not speak to the MOD. The DTI contacted Dr Bayliss of Walter Somers and then asked Mr Weir to speak to the company. This he did on two occasions, but had not enough information to reach a conclusion about whether the tubes might be civil or military, although it was acknowledged that he was informed of the involvement of Gerald Bull's company, SRC.[8]

## The case put by Sir Hal Miller and the companies

Sir Hal disputed this version of events when he gave evidence to the Scott Inquiry, and contemporaneous notes of his conversations with the Department support his contentions.[9] The notes recorded *inter alia* the number of tubes, their dimensions and the quality of the steel, as well as SRC's involvement and the telephone numbers of Weir, the MOD metallurgist, and Steadman, the Director of the Export Licensing Branch. Sir Hal asserted that he was clearly conveying to both the DTI and the MOD information that the company suspected that the end use was military and this was why, when Walter Somers was in a parlous financial state, it was drawing their suspicions of a valuable contract to the attention of government. The company was suspicious because they knew both that the tubes were destined for Iraq and that the specification of the steel had been changed to one, H65, that was not appropriate to petrochemical uses but was to military applications. In fact, the specification is a Royal Ordnance one. Dr Bayliss had made gun barrels for the MOD. Forgemasters had also made guns and were, at the time, a wholly owned subsidiary of British Steel, in turn then a nationalized industry.[10]

Sir Hal asserted that he passed on the company's offer to complete or to drop the contract, or to allow it to be traced. He described the contract as comprising 'missile tubes',[11] and reported the involvement of an Iraqi called Assawi. Sir Hal further asserted that he was rung up by a man calling himself 'Anderson' from the MOD. When Sir Hal related what the company had told him, he was told by Anderson that, 'This confirms everything we know.'[12]

## The Scott Report

The account produced by Sir Richard Scott supports that given by Sir Hal Miller and Dr Bayliss save in two respects: Sir Hal's contacts in 1988 with both the MOD and a member of the SIS/Security Service were disputed.

Following the telephone call from Sir Hal Miller, the Director of the ELB, Mr Steadman, alerted Mr C3 in the SIS 'in view of the involvement of a Space Research Corporation'. He also contacted Mr JJ in the DIS. The matter was passed to Mr Weir who made a note of: SRC's 'work in the armament design/development area'; the involvement of both Walter Somers and Sheffield Forgemasters; and the nature of the steel.[13] Weir knew the nature of SRC and the link with Sheffield Forgemasters in late 1987 because of an ELA referred to him then. At that time, he had also provided the SIS with a possible 'lead' within SRC, a former colleague, a 'brilliant gun designer' who had worked for the MOD's Royal Artillery Research and Development Establishment (RARDE).[14]

Weir made notes of his conversation with Dr Bayliss on 16 June 1988. He recorded the latter's concern that Walter Somers was being asked to manufacture missile tubes and the specifications of the tubes.[15] Sir Richard Scott did not accept Weir's suggestion, made to both the Trade and Industry Committee and the Inquiry, that Dr Bayliss attempted to mislead him.[16]

Weir wrote up his conclusions: while there was no definitive answer to the question of final application, he hypothesized a number of military uses. In a second conversation with Dr Bayliss, Weir noted the fact of their agreement that it was highly unlikely that the tubes were to be used for petrochemical purposes and the Iraqi connection.[17] After discussion with Mr G in the DIS, Weir made an addendum to his notes: 'The possibility of use for nuclear studies should not be ignored.'[18]

Sir Richard Scott found no documentary evidence that Weir's notes were either disseminated, or discussed in either the MODWG, which dealt with exports to Iran and Iraq, or in the REU, the forum for intelligence sharing. Mr G observed that his desk in the DIS was 'not equipped to deal adequately with the problems of exports to (non-Soviet bloc) countries of equipment for their sole internal use', but thought his colleagues, Mr J and Mr N3, both members of the REU and MODWG, knew of the tubes.[19] In spite of Weir's usual office practice of sending his assessments to the person who made the original request, in this case Mr JJ in the DIS, Sir Richard found no evidence that he did in fact do so. Weir had a copy of his notes, but not the original, in his possession. Nor did the Inquiry find any documentary evidence of the advice Weir thought he would have given to Steadman in the DTI, namely that he had no information that the tubes were of military significance.[20] Weir's notes were at variance with this advice to the DTI.

From July 1988, 'Mr Weir appears to have begun a series of further contacts with the Security Service and SIS about the Walter Somers forgings.' A Security Service Note, dated 5 July 1988, records Weir as having spoken to Miss M in the SIS and Mr F and Miss H in the Security Service about SRC's specifications as suggesting either 'pressure vessels' or 'gun barrels'. The record ends with an historical note about the use of large calibre guns. The 5 July Note is at variance with the notes Weir said he made in June as being less circumspect about a military application. Neither Miss H nor Mr F raised the matter in the REU even though Miss H was a regular attender and Mr F received the REU minutes.[21] However, Mr F was concerned with terrorist-related arms trafficking and Miss H with illegal exports to COCOM proscribed destinations where hostile intelligence services were involved.

Meanwhile, Sheffield Forgemasters contacted the DTI on 17 June 1988 to ask about licences and were asked to send details of the contract. They did so the following day, enclosing a drawing with 'comprehensive information'. Although the DTI could not find the drawing on their files, the Trade and Industry Committee concluded that one had been sent.[22] Mr Draper, who dealt with Sheffield Forgemasters, sent the information to Mr Jacob of the Industrial Metals and Minerals Division of the DTI, who in turn advised that no licence was required. In August, Walter Somers telexed the DTI, identifying the link between the company and Sheffield Forgemasters. Again, Draper contacted Jacob who gave the same advice as in the Sheffield Forgemasters inquiry. Walter Somers offered to send drawings to the DTI but were told by Mr Draper, who had seen the Sheffield Forgemasters' drawings, that there was no need to do so.[23]

In 1990, MOD experts, having seen the drawings sent by Sheffield Forgemasters to the DTI in June 1988, concluded that they would have been immediately suspicious and investigation would have raised the possibility of 'elements of a very large gun'.[24] However, since Steadman had not briefed the ELB, nor informed the REU, neither Mr Draper, nor others in the DTI, knew of the official contacts and suspicions. Steadman had bypassed normal procedures by not making a formal inquiry through the DESS.[25]

In November 1988, Mr Weir discussed the events of the previous June with Mr I in the Security Service, which the latter then noted to Mr F and Miss H. Mr I reported that 'Weir has various theories believing fundamentally that (the tubes) are some form of weapon or weapons testing' for Iraq.[26] None of the Security Service officers took the matter further because it was not directly relevant to their fields of interest. Nor did they report it to the REU, which both Mr F and Miss H attended days after receipt of the Minute, because they assumed others in the MOD or the SIS would deal with the information.[27] Such compartmentalization has now been eradicated with the counter-proliferation section in the Security Service. It examines information regardless of the involvement of hostile intelligence agencies.[28]

In mid-1989, Weir approached Miss M to say he had heard nothing from the SIS and to ask if she had passed on to Mr C5 the information he had provided in 1987 about the former RARDE gun designer working for SRC. Miss M had done so but spoke to Mr C3, Mr C5's successor, about the 'long-range artillery'. Miss M said that Mr C3 agreed to take the information forward, although Mr C3 had no recollection of Miss M speaking to him. He thought it probable that he passed the information to Mr Q who had responsibility for conventional defence equipment. Mr Q did not recollect the contact. Miss M later checked with Weir that he had been contacted. He had not; so, in Mr C3's absence, she spoke to Mr C2. Mr C2 did speak to Weir, but does not remember Miss M telling him to do so.[29] The two men had no further meetings because Mr Q then took the investigation forward.

Mr Q had joined the SIS in April 1989 from the DIS. He had seen neither the notes Weir made in June 1988 nor the Security Service Notes of 5 July and 4 November 1988. Had he seen any one of them he 'would have reacted'.[30] Sir Richard Scott believed that it was Mr Q's pursuit of the evidence in late 1989 and early 1990 which lead to Customs' seizure of the tubes.

When, in June 1989, Walter Somers received another order for tubes for Iraq, the new Managing Director of the company, Peter Mitchell, again contacted Sir Hal Miller who again spoke to the MOD as well as the DTI in June and in August. The MOD asserted that it had no records of Sir Hal's contacts in 1989.[31] Sir Hal stated that when he spoke again to Walter Somers following these contacts, he was told that the orders had been approved by the DTI.[32] Sir Richard Scott accepted the account given by Sir Hal and Peter Mitchell of Walter Somers.[33]

It was in the autumn of 1989 that the final clues to the supergun's existence began to emerge. The first was a contract between the British company Hadland and Iraq, for short duration flash ballistics shadowgraph equipment, which was connected to a military related project, 'PC2-BABL' and a Dr Assawi, an Iraqi first mentioned in the initial contacts between Walter Somers and the DTI and the MOD. The project was connected to SRC's experimental satellite launchers. Hadland's involvement in a new project codenamed 'Babylon' was reported to the WGIP in November 1989.[34]

The second clue came from the British company, Astra, which had bought PRB, the Belgian company manufacturing the propellant for the supergun on behalf of SRC. Astra personnel reported an 'unusual contract' for Iraq via Jordan to both the SIS and to the MOD in September and October 1989. These reports resulted in the DTI undertaking 'to report any applications that were received', although, because 'there was no evidence of a breach of UK law', no investigation of British companies' involvement was undertaken by the DTI. The SIS recommended that a démarche be made to the Belgian Government to prevent the export of PRB propellant. This was issued on 21 December 1989.[35] Mr Primrose, for the MOD, and Mr Q, for the SIS, agreed that all relevant information 'should be reported to Ministers as a matter of urgency'. However, this was not done.[36]

The 'third straw in the wind' for Mr Q was information from Paul Grecian, the Managing Director of Ordnance Technologies. Mr Grecian reported a long-range gun, Babylon, being developed for the Iraqis by SRC.[37] On 6 October 1989, Mr C2 of the SIS prepared a briefing note for a Special Branch meeting with Paul Grecian. The SIS note referred to the MOD's 'indications of an Iraqi interest in long-range artillery weapons' and SRC's attempt in June 1988 to acquire 'artillery gun barrels' from Walter Somers for Iraq's PC2 project.[38]

Sir Richard Scott noted that, 'It is clear from the information in the 6 October briefing note that SIS had been in contact with Mr Weir', but that it remained unclear which official(s) in the MOD had formed the concrete opinion that the tubes were for artillery gun barrels. Mr C2 thought if he had got that information from 'a SIS source', while Mr Weir denied being the originator.[39]

On 5 December 1989, the SIS weekly digest of intelligence noted Iraq's development of an 'ultra long-range gun'. The digest was sent to the Prime Minister and the Foreign Secretary, but not the Secretary of State for Defence.[40] The tubes were eventually seized in April 1990 because information from Astra, Paul Grecian, and David James, Chairman of Eagle Trust, which owned Walter Somers, helped Mr Q to put the pieces of the jigsaw together.[41]

## Unresolved issues

### The Government's credibility

In Sir Richard Scott's report, the documentary evidence is relied upon to establish *each stage* of the story, so that the parts become more than the sum. No pattern is sought, no case constructed, no explanation for the totality of events reached for. The Scott Report relied upon government witnesses and official documentary evidence in producing an account of the supergun affair. However, given that the Government's credibility has been severely undermined, uncertainty hangs over the reliability of the narrative produced by Sir Richard. Clearly, as the Scott Report found, the Government had been less than truthful about the events surrounding the supergun.[42] Sir Richard allowed the blame to fall on officials for their inadequate briefing of Ministers in answering questions; but this is based upon an assumption that such briefings reflect the whole truth as told to the Minister rather than what civil servants know Ministers would wish to tell.

Moreover, the Government was less than willing to allow certain witnesses to give evidence to the Trade and Industry Committee's investigation. This is also suggestive of a continuing effort to suppress evidence of the supergun affair, as was the behaviour of a member of the Government who rang Gerald James, who had informed the authorities about PRB's involvement, to warn him 'not to go over the top' in his evidence.[43] The Chair of the Committee defended meeting the former project manager of the supergun, Christopher Cowley, the night before he gave evidence by saying that, 'I just did not want him to feel there was any pressure to say things that he was not sure of.'[44]

Although the Scott Inquiry could find 'no contemporaneous record' of why it was thought inappropriate to proceed against Christopher Cowley, Sir Richard accepted Customs' explanation that, given the dropping of charges against the companies involved, it was thought unfair to prosecute Cowley when his involvement had ceased at the time of the exports.[45] More significantly, the prosecution of Peter Mitchell of Walter Somers was dropped because Customs could not satisfy the evidential sufficiency criteria in the Code for Crown Prosecutors. In short, when subjected to the sort of forensic examination conducted by Defence Counsel, the case would have fallen apart.

### Non-government witnesses

Participants in the defence trade with Iraq argued that the supergun was the subject of 'common gossip' in industrial and political circles in Baghdad.[46] Walter Somers' Managing Director, Peter Mitchell, told Sir Hal Miller that when he saw the British Defence Attaché at the Baghdad Fair in 1989, the Attaché knew about Project Babylon.[47] Belgian government sources claim to have told Downing Street about the supergun in May 1988.[48]

Chris Cowley, the supergun's project manager, asserted to the Trade and Industry Committee that the intelligence agencies were told of Project Babylon before the contract was signed and that they were kept informed of progress, including the use of British companies, by Bull, who in turn sought their approval. SRC officials, Cowley asserted, also briefed the State Department in Washington in February and March 1988.[49]

Cowley submitted evidence to the Committee which it decided not to publish, or to refer to, in the report after the Chair used his casting vote to settle the matter. Phythian argued that:

There exist a number of faxes and letters in Cowley's unpublished testimony which give the impression that contact with people in or around the British Government was an ongoing process.[50]

The Scott Report referred to one such fax which supports Cowley's assertion, and that of Dr Bull's son, Michel, that Bull kept the British Government 'well informed' about the supergun.[51] However, Sir Richard concluded that, since Bull was prepared, at the inquiry stage, to mislead the British companies involved in the supergun about the true nature of the project, he was likewise prepared to deceive others about his contacts with the authorities. The Inquiry found no documentary evidence of contacts between Bull and British intelligence.[52]

In a letter to the Clerk to the Trade and Industry Committee, Cowley asserted that when he attended meetings for SRC in China, George Wong, a director of Rothschild Bank, was always present. Such meetings were concerned with technical, not financial, matters. Bull had told Cowley that Wong acted as a conduit to British Intelligence. Cowley attested that, 'I fail to see what role Wong played in these discussions if it was not to report back to some higher authority.' Cowley claimed that Bull talked often to Wong about Project Babylon.[53] Wong denied such discussions took place and that he was an agent for the SIS.[54]

However, it is possible that Bull kept the intelligence agencies informed through other means or other people. The DTI admitted in 1992 to solicitors acting for a former SRC engineer that government meetings and correspondence with SRC were 'numerous and confidential'. A former company executive said that the meetings concerned 'details of most of the work the company was doing – including the Iraqi work and briefings on British sub-contractors'. The meetings began in at least 1987 and took place as often as once a month in London.[55] Many of SRC's mainly British workforce were former employees of the MOD's RARDE.[56] The Scott Report established that, in 1987, the SIS was looking for a 'lead' in SRC and that a former RARDE gun designer had been identified by Weir. It is also possible that at about the time of the initial inquiries, the SIS recruited an agent from one of the companies manufacturing the tubes for Iraq. Sir Hal Miller thought this a strong possibility and thought he knew who the recruits were.[57]

The Belgian authorities' assertion that they told the British Government about the supergun in May 1988 would cohere with Sir Hal Miller's assertion that he was told by British intelligence that his information confirmed what they already knew. Sir Hal Miller recalls speaking to 'Anderson' who he understood to be a member of the intelligence services. Anderson had told him not only that Sir Hal's information confirmed what was already known, but also that 'We'll run a little game plan.'[58]

Sir Richard Scott found this to be implausible for three reasons. First, Sir Hal asserted in two instances that Weir had put him in touch with Anderson and in another instance that Weir was appointed as 'go-between' by Anderson. Second, Sir Richard found the appointment of such a go-between unnecessary. Third, Sir Hal did not mention Anderson as saying that Sir Hal had confirmed everything they knew in his correspondence with the Prime Minister and the Cabinet Secretary in April 1990.[59] However, there is nothing inconsistent in Sir Hal being put in touch with Anderson and then Weir subsequently acting as go-between. Nor is it implausible that Weir should act as go-between. Sir Hal's failure to mention Anderson to the Prime Minister or Cabinet Secretary was entirely consistent with his general behaviour in not wanting to say more than was necessary to protect the companies from prosecution. Moreover, Sir Hal's credibility as a witness should also be considered.

There is no discernible incentive for Sir Hal, a lifelong Conservative and a former member of the Government, to make accusations which reflect badly upon the administration. He held to his silence on the matter until the Scott Inquiry, save to warn that he would give evidence if the companies found themselves in court. He kept silent to avoid embarrassing his party and his Government. In any case, further evidence of his credibility exists in his frequent correspondence with, *inter alia,* the Prime Minister, the Cabinet Secretary and the Attorney General when prosecutions against Walter Somers and Sheffield Forgemasters were under consideration.[60]

His version of events, after the smearing of his conduct and character, has been found to be more truthful than the official account. Furthermore, Mr L2, a senior DIS official, argued that Sir Hal could have spoken to someone from 'an SIS branch located within MOD and whose existence is not normally acknowledged'.[61]

Apart from the evidence of non-government witnesses, the implausibility of the supposed actions of officials in the account given in the Scott Report also lends support to the hypothesis that the SIS began monitoring the Babylon contracts, or using the companies involved as a point of intelligence access to the Iraqi procurement network.

## The DTI's actions

The Director of the Export Licensing Branch, Mr Steadman, attended both the IDC and the REU, the latter body having been set up to share intelligence on just the sort of illicit procurement that the contracts with Walter Somers and Sheffield Forgemasters represented. The Director had had a long involvement in export licensing for Iran and Iraq. While he was not concerned about Iraq's procurement of technology for conventional arms manufacture, he was very much aware of the SIS's monitoring of the Iraqi procurement effort.

Steadman's superior, speaking on another matter, said that suspicious export licence applications would be passed by officers to Steadman, 'so he passed them on to SIS'.[62] The Director of the ELB is thus established as having a close relationship with the SIS. The Director did contact both the SIS and the DIS in June 1988. Given that this demonstrates his awareness of the importance of the information he had been given on behalf of Walter Somers, there is no obvious explanation of his failure to pursue the matter through the DESS, the established channel, or to report the information either to the REU or to other members of the ELB. Steadman did not give evidence to the Trade and Industry Committee.

As for the behaviour of Draper, the DTI official who told the companies that licences were not required, the Treasury Counsel's Opinion of the prosecution case against Walter Somers made the following point. Although Draper had said that it was not 'obvious' to him from the drawings sent by Sheffield Forgemasters that 'the goods were for anything other than the petrochemical industry', Customs' experts, including a metallurgist with expertise in the petrochemical field, had concluded that it must have been obvious from that information that the tubes 'had no application in the petrochemical industry and were parts of a barrel'.[63]

When Walter Somers inquired about an export licence in August 1988, Sir Richard Scott raised the question of why Draper failed to accept the company's offer of drawings. Albeit that the inquiry was for the same contract as the previous query by Forgemasters, it was nonetheless from a different company making a separate rating request.[64] Two further matters also remain puzzling: that the drawing sent to the DTI could not subsequently be found; and that Draper merely *discussed* the information from the company with Mr Jacob of the Industrial Metals and Minerals Division of the DTI and did not forward the drawings.[65]

Finally, as the Treasury Counsel asked: 'Was Steadman unaware of the application? If he was not, why did he not investigate it fully and possibly contact the MOD in the light of the earlier enquiry?'[66]

### The DIS's actions

The DIS were assiduous in pursuing issues of Iraqi procurement and, unlike the DTI, were concerned about Iraq's capabilities in conventional weaponry. They were active opponents of the export of machine tools to Iraq throughout 1988 and beyond. In April 1988, the DIS requested from the SIS 'details on the precise nature of armaments production in Iraq'.[67]

In June 1988, Steadman contacted Mr JJ in the DIS who was head of a section concerned with monitoring and assessment of certain development and production programmes for all types of missile and space programmes. In a note to Weir, Mr JJ recorded that the size of the tubes made them unlikely, but not impossible, as Third World missiles, but that 'the material may have other military uses'. He asked Weir to contact him 'a.s.a.p. tomorrow morning'. However, the two did not discuss the matter further.[68] Again, Treasury Counsel's Opinion can make the point that: 'these orders . . . were . . . among the largest forging contracts since 1945. Added to this was the political reality that Iraq was at war'.[69]

### The SIS's actions

By June 1988, the SIS knew that Iraq had used indigenous medium-range ballistic missiles against Iran and was procuring technology for its defence industry by using British companies. Furthermore, the SIS was actively monitoring Iraq's procurement network. It had been approached, by another country's diplomats, for information about Gerald Bull and SRC. The response, in April 1988, showed that the SIS were already aware of his and his company's activities.[70] The SIS was seeking 'leads' within SRC.

Steadman approached Mr C3 in the SIS after the initial contact on behalf of Walter Somers. Mr C3 was responsible for writing reports for Whitehall on Iraqi procurement;[71] indeed, it was his reports which led to the DIS's request for details on Iraqi defence manufacturing capabilities. Mr C3 attended a meeting of seven intelligence officers to discuss Iraq on 1 March 1988, three months before the contacts with Walter Somers. The meeting noted the 'high interest on further infor-

THE SUPERGUN ◆ 163

mation on attempts to export armaments or the means to produce armaments' by Iraq.[72]

In intelligence terms, the Bayliss and Miller contacts provided two related opportunities: to gather intelligence on SRC through the two British companies; and to learn more about Iraqi procurement, especially since Dr Bayliss had both provided the name and passport number of Abdul Shibib and mentioned an Iraqi, Assawi, who was to sign the contract. Moreover, the company had made the offer to allow the order to be traced.

### Bill Weir

Mr Weir said that he had no knowledge of Iraqi procurement and his function was to assess materials, not weapons systems.[73] However, he would have been aware by 1988, as any newspaper reader, of Iraq's use of chemical weapons and the country's development of medium-range missiles. It was an ELA for a special high-temperature nickel alloy to be used as a lining for a 155mm gun barrel which first brought SRC and Sheffield Forgemasters to his attention in 1987; and Mr Weir's lack of expertise *vis à vis* weapons systems was not an impediment to his joining the UN Inspection Team that visited Iraq after the 1991 war.[74] Weir's position or section has never been identified, which tends to suggest that he was connected to the intelligence agencies.

There are four puzzles in Weir's actions in 1988 and 1989. First, given what he said subsequently to the SIS and the Security Service, it is odd that he should tell the DTI that he had no information to suggest a military end use. Second, it is equally odd that he did not submit his opinion in writing. Third, it remains to be seen why he did not wait until drawings became available to make an assessment. Fourth, he seems always to have pursued the issue with the intelligence agencies and never with colleagues in the MOD.

### Other evidence

According to the narrative produced by Sir Richard Scott, there are at least twelve officials who decided not to take forward the fact that a British company believed it was making missile tubes for Iraq. Given that this is a bright and dedicated group of people, the logical explanation for this succession of omissions is that the matter was in hand elsewhere. Sir Richard found no documentary evidence of the matter being discussed in the REU, but as Mr F argued:

That something was not mentioned at the REU, or in its minutes, was no assurance it had not been discussed outside the REU, or in the REU's margins. That something did not figure in the minutes was no guarantee it had not been mentioned at a meeting.[75]

Other circumstantial evidence points to an ongoing intelligence operation. For example, Mr C2 insisted that the SIS held information that 'in June 1988, SRC had been involved in an attempt to acquire Walter Somers tubes "probably intended for use as gun-barrels"'.[76] This is supported by the contents of the SIS briefing note of 6 October 1989.

By December 1988, according to Mr C3, access to the proliferation activities of the Iraqi network did not rely heavily upon Matrix Churchill.[77] If employees of SRC or the British companies contracted to produce parts of the supergun had become agents for the SIS, it would explain why, in the summer of 1989, when the DTI and the MOD were again contacted by Sir Hal Miller about a further order for tubes for Iraq, no action was taken. It would also explain the export licence given to another UK firm, Hadland Photonics, which supplied ballistics research equipment. The Trade and Industry Committee reported that the latter company supplied some equipment to Iraq which did not require a licence and was refused one export licence in October 1990.[78] However, the MODWG study of British assistance to the Iraqi arms industry highlighted Hadland's supply, between October 1989 and October 1990, of ballistics research equipment worth over £1 million. Two comments accompanied the listing: 'query MTCR' and 'should be 1A'. The DTI had classified the equipment as industrial when it should have been military.[79] The export was picked up by the intelligence agencies since it was reported to the WGIP.

By November 1988, senior FCO and SIS officials had decided to make Iraq the SIS's top priority because of the perception that Iraq was seeking 'a poor man's nuke'. It is possible that it was the British forgings which lead to this conclusion. Again, an Intelligence Report dated October-November 1989 and headed 'Iraq: Project "Babylon" to Develop the Technology for Hypervelocity Gun with Extreme Range Capability', provides other possible clues (see Appendix B). The contents of the report suggest both that Dr Bull was well known to the SIS and, since the Report contains much assessment, that the raw intelligence on the project was received at a much earlier date.

The action by Customs prevented the export of the final consignment of eight of the total of fifty-two tubes for the supergun. PRB was similarly prevented from fulfilling its contract with Iraq after an explosion at PRB in December 1989, labelled an act of sabotage by Belgian police. The SIS had told the parent company to continue with the export of the trial batch of supergun propellant after Astra informed them of the PRB contract in September.[80] Gerald James maintained that Astra would not have bought PRB in 1989 had the company not been strongly encouraged to do so by the Chairman of Astra's main shareholder, the 3i investment bank, Sir John Cuckney, and by a non-executive Director of Astra, Stephan Kock. Both men had connections with the SIS. Through Astra's purchase of PRB, the intelligence agencies were able to gain access to the Belgian company.

Those civil servants and Ministers who appeared before the Trade and Industry Committee argued that Ministers were not informed of the contacts with the various companies.[81] This might well be true. In fulfilling the requirements of users, SIS might not have had to report every detail of its intelligence gathering operations to Ministers. But, as the intelligence picture developed and the strategic implications of the supergun became apparent, Ministers were made aware of the project although not, apparently, the involvement of British companies. However, so much of Whitehall's business, especially in sensitive matters, is conducted by word of mouth. It can be assumed that, at the very least, the Foreign Secretary and the Prime Minister would have been informed; the first because of his responsibility for the SIS, and the second because of both her general interest in intelligence matters, attested by her Private Secretary,[82] and her specific interest in Iraq's activities. As a member of the Trade and Industry Committee said, it was 'inconceivable that civil servants would not have informed Ministers at the highest level'.[83]

As in the Matrix Churchill case, successive orders for the tubes might have been quietly permitted because the SIS had told officials and Ministers that such a course of action was necessary to protect a source of intelligence. Equally, as in the case of the various machine tools ELAs, Ministers might well have been made aware of both the nature of the exports in question and the issue of source protection, but the decision to allow the export would have been theirs to make. In the case of machine tools, the fact of their suspected military end use meant that officials referred all decisions to their Ministers: they did not decide for themselves.

While this is, of course, speculative, such a scenario is more realistic than the coincidence of a dozen officials failing to act on the information that British companies were making missile tubes or long-range artillery for Iraq. As for the Scott Inquiry, it is quite plausible that the Vice-Chancellor was allowed to chase the paper trail of Weir's inquiries as to the progress of the SIS operation and was given access to the final stages of the intelligence *gathering* performed by Mr Q, either in tandem or independently of the existing operation. The history of the Scott Inquiry is the history of the struggle to wrest documents from the Executive. By definition, it is not known which documents Sir Richard failed to get.

# 8

# Machine tools

## Export licensing of machine tools

The case of machine tools exports to Iraq can be contrasted with that of the supergun in order to better understand the UK's considerations in exporting dual-use equipment. Again, a fairly detailed narrative is necessary in order to establish that the licensing of machine tools represented a series of political decisions rather than failures in control mechanisms.

Between 1987 and 1990, export licences were issued for four batches of Matrix Churchill and other companies' machine tools for Iraq. For each decision, intelligence was available about the general intentions of Iraq with respect to procurement and the specific military end use of the goods in question. Officials and Ministers explained these four licensing decisions by reference to three factors: the need to protect a source of intelligence; failures in the retrieval and dissemination of intelligence; and the unreliability of intelligence. In short, the export of machine tools was explained by reference to implementation, not to policy. However, while an examination of the narrative reveals that there are elements of truth in these assertions, the main reason for allowing exports of equipment for Iraq's military industry lies in the UK's pursuit of influence and trade.

An intelligence report, based on information provided by a sales director of the Iraqi-owned Matrix Churchill, was circulated in November 1987. The report said that machine tools, recently licensed but not yet shipped by Matrix Churchill and other companies, were to be used in the manufacture of shells and missiles at two facilities in Iraq, Nassr and Hutteen, the latter also known as Iskandariyah. Licences were suspended while officials considered what advice to give Ministers. The SIS approached senior officials in the Foreign and Commonwealth Office, the Department of Trade and Industry and the Ministry of Defence to argue that, in order to protect their source, they would prefer 'export licences refused only for future suspect business'.[1]

The licensing of the second batch of machine tools in early 1989, for the *Ababel* rocket plant, the ABA project, cannot be as easily ascribed to the need to protect an intelligence source because the SIS had developed other sources by this point and collateral for the original intelligence had come in various forms throughout 1988. The first was a letter from an employee of Matrix Churchill which said that machine tools were to be used to make shell cases. The second was a telephone call from a managing director about arms manufacturing at Hutteen. Both pieces of intelligence were circulated by the FCO, to which they were first sent, to other Departments. In April, the Military Attaché in Baghdad telexed the DIS with intelligence about Nassr and Hutteen. He could not, he said, 'see any reason for granting export licences for anything connected with these two areas; unless you are devious enough to wish to gain a unique entrée!'[2]

In December 1988, Dr Safaa al-Habobi, known to be the head of the establishment for which the machine tools were destined, was implicated in attempts to procure equipment for the development of gas centrifuge technology for uranium enrichment. However, this evidence of 'expansion of activities of the procurement network into the nuclear proliferation field' was used by the SIS to emphasise the importance of source protection and thus the granting of licences.[3] This was in spite of the fact that these particular machine tools were 'essential for the production of nuclear weapons'.[4] The DIS, however, took a different view from the SIS, arguing that this intelligence meant that 'the line must be held until such time as this has been evaluated as we could be guilty of a far greater folly if we give ground now'.[5]

In a minute to the DESS, copied to the SIS, the DIS highlighted its concerns about machine tools exports. They noted that the SIS was 'adamant' that their original intelligence could not be used in isolation but argued that the DIS's assessment of Hutteen as 'the main ammunition manufacturing plant in Iraq' was not wholly dependent upon the SIS information. Furthermore, while the assessment of Nassr had relied solely upon the SIS source, collateral had now been received in the shape of an application for preliminary clearance to export a moulding machine to produce polyurethane internal filler parts for missiles. The DIS, therefore, thought it feasible to refuse the licences for machine tools without compromising the SIS source.[6] The DIS lost the argument. In spite of subsequent independent intelligence, licences were issued in February 1989, argued participants, to protect the SIS's source.

The principal explanations offered by officials and Ministers for allowing the export of the third and fourth batches of machine tools revolve less around the issue of source protection and more around problems of disseminating and retrieving intelligence as well as questions of its reliability.

The third batch of machine tools for Iraq comprised: twenty-four machining centres and other lathes for the manufacture of fuses at a large munitions complex built by Industrias Cardoen at Al Fao General Establishment; and machine tools for Project 1728, a programme for extending the range of *Scud-B* missiles. Although both sets of machine tools were licensed as one batch on 2 November 1989, they are discussed separately because of the difference in the availability of intelligence.

Consideration of the Cardoen machine tools, as well as those for Project 1728, began in January 1989. Although specific intelligence about the contract was not received by Departments until October, the company had not hidden Cardoen's involvement – indeed the DTI queried the company's involvement – and the Nassr complex was known to be the destination. Moreover, Cardoen's supply of munitions to Iraq was first reported in the press in June 1984.[7] The Chilean manufacturer's business was, in short, well known. The ECGD had established that much in 1988 merely by reference to press clippings,[8] and the Royal Ordnance Factory had conducted a Hazard Assessment of Cardoen's Bomblet Plant in 1987.[9]

On 13 October 1989, GCHQ circulated a report to the FCO, MOD, DTI and Customs and Excise. Industrias Cardoen was to turn over to Iraq a munitions factory and an associated cluster bomb plant; and a UK firm was to supply twenty-four machining centres for various types of fuses between August 1989 and March 1990.[10] It was not until the REU meeting on 24 November, six weeks later, that the DTI raised the report. GCHQ, the minutes recorded, would supply the company name. It should be noted that the DTI did not seek the company's name and that neither the MOD nor the FCO raised the issue. At the REU meeting on 8 December, one week later, the company's name was given as Matrix Churchill and was written in the minutes in the upper case.[11]

One has to ask first why officials failed to connect the twenty-four British machining centres in the report from GCHQ with the twenty-four machining centres with which they had been dealing over the last ten months. Steadman, the Director of the ELB, who received the report, had investigated Cardoen's involvement in April and had written four submissions on the machine tools in that year.[12] The FCO's Middle East Department, which also received the report, had written five submissions on this batch of machine tools from January to October 1989. Second, it is curious that it should take six weeks to raise the report, three weeks after the machine tools were licensed. Third, the question arises as to why, when the name was revealed, the export licences were not revoked. The machine tools were not due to be shipped until January 1990.

Officials who attended the REU when the company's name was given explained that, because they had not seen the original report, it meant nothing to them and therefore they took no action. However, it is impossible to believe that the three officials at that meeting who had, in that month, written submissions for their respective Ministers upon the issue of Matrix Churchill export licences and had been involved in a long process of consideration, did not make a connection.

On the remaining machine tools, i.e. those destined for Project 1728, intelligence of a very specific nature was available from at least June 1989. Officials argued that they had neither sent intelligence reports to Ministers nor reflected the content of them in their many submissions to Ministers because they had not received those reports at the appropriate time. Past intelligence reports had been forgotten since it was not permitted to take a note of them.[13] However, since there were so many reports on Project 1728 and the content of them was discussed and recorded in meetings, a collective failure of memory will not suffice in explanation.

On 23 June 1989, the WGIP noted that for two of Matrix Churchill's ELAs, 'the end user was known to be the Iraqi missile programme'. The Director of the ELB who dealt with the applications was a member of the WGIP, as was the Chair of the MODWG who dealt with the Matrix Churchill case in the MOD. The FCO's Middle East Department also sent a representative to the WGIP.[14] In July and September 1989, four separate intelligence reports said that Project 1728, clearly stated as the destination of machine tools on the two Matrix Churchill licence applications, was a large ballistic missile programme involving UK companies.[15] In September, GCHQ identified Dr al-Habobi as being involved in the missile project.[16] On the question of Iraq's intentions, one of the reports stated that the priority was military rather than civil.[17]

Apart from the Director of the ELB's attendance at the WGIP which had given the end use of some of the Matrix Churchill machine tools, one can trace the DTI's grasp of the information in other ways. As a consequence of the July reports, the DTI telephoned Matrix Churchill requesting clarification of Project 1728 and the machine tools' end use. The reply simply stated that the Project was part of 'a number that Nassr are embarking on in order to use their suitable machine capacities'.[18] This seems to have satisfied the DTI's inquiry. An internal DTI memorandum in September noted that the machine tools 'are destined for military manufacturing purposes'.[19]

For the MOD, the Chair of the MODWG attended the WGIP meeting; he also saw the intelligence reports.[20] In the FCO, the end use of the machine tools was widely circulated, not just because of the WGIP and the intelligence reports, but because, as a result of these, the Science Energy and Nuclear Department (SEND) of the FCO had become involved in the decision-making process. The SEND had wanted to turn down the applications but had relented when the DTI undertook to use the opportunity provided by the ELAs to elicit information from Matrix Churchill about Project 1728. This did not happen.[21] In September, the SEND wrote a minute, copied to the Minister, about the company's ELAs for Iraq's missile programme, recommending refusal of the licences.[22] Finally, it should be noted that officials in all three Departments made submissions to their Ministers shortly after the intelligence reports of the autumn: the DTI on 8, 20 and 26 September; the MOD on 22 September; and the FCO on 26 September.

Machine tools destined for both the missile project and the munitions plant were licensed, the FCO argued, because the Department, which had wanted to refuse the applications was forced to concede to the DTI and the MOD when the SIS changed its view of Iraqi intentions. According to the relevant FCO submission from October 1989:

> Our friends have said that they believe that the lathes may not, at any rate initially, be used for the direct manufacture of munitions or for nuclear applications. They are inclined to believe statements by Habobi ... that his organisation is now dedicated solely to the post-war reconstruction of Iraq.[23]

Although the SIS did not circulate a written assessment in the terms above, the submission writer asserted that he 'would have had to have' checked the accuracy of the briefing with the SIS.[24] The submission was copied to the Permanent Under-Secretary's Department, thereby ensuring it reached the SIS since it is that department in the FCO which is responsible for the SIS. There is no evidence that it was at any stage challenged by the intelligence agency even though it did not represent its opinion.[25] The SIS would not have been inclined to believe the statements of a man known to be engaged in trying to buy components for nuclear weapons and identified as being involved in Iraq's missile project: such statements, in any case, contradicted some very precise intelligence, received just weeks earlier, about the destination of Matrix Churchill machine tools.

By this time, the SIS was not arguing that its intelligence source had to be protected by allowing Matrix Churchill machine tools to continue being exported, perhaps because, by then, that position had become untenable. Sir Richard did not think that the temporary change in the SIS's assessment was done 'in pursuit of any covert SIS policy';[26] rather, the SIS was unconcerned by Iraq's procurement of conventional weapons.[27] An alternative explanation is that the SIS sought, by changing its intelligence assessment, to maintain its Matrix Churchill source by keeping the agent's company in business. In early 1989, the SIS began to share intelligence on Iraq with the CIA.[28] According to Paul Henderson, the Managing Director of Matrix Churchill and an SIS agent, the Americans had lost their intelligence access in Iraq.[29] The SIS might have had their own organizational motive for changing the assessment; or they might have been indulging in the same apparent wishful thinking that had beset decision-makers about Iraq's switch from military to civil procurement after the ceasefire with Iran.

No official challenged the SIS's new assessment in late 1989 even though it contradicted all the intelligence of that year. Nor did any Minister because none was, it was argued, in a position to do so. The Foreign Office Minister, William Waldegrave, thought that the 'essence' of the specific intelligence relating to Project 1728 should have been reflected in the submission to him.[30] The MinDP, Alan Clark, argued that had he seen the precise intelligence – 'one fifth of what is here' – his attitude to exporting machine tools to Iraq would have been different.[31]

## Policy and implementation

The case of the final batch of machine tools licensed in 1990 undermines the claims made by officials and Ministers that all four decisions were made because of the operation of one or more of the following factors: the necessity to protect an intelligence source; problematical retrieval and dissemination of intelligence; and the inherent unreliability of intelligence. That fourth decision also undermines Ministers' claims that they would have acted differently had they known of the precise intelligence. Before discussing the 1990 decision, it is useful to assess the strength of the above claims in the process of decision-making for the previous three sets of machine tools.

In July 1990, the SIS minuted the Cabinet Office to dispute that source protection had been the primary issue in export decision-making about machine tools.[32] This assertion is supported by minutes of the REU meeting on 22 January 1988 when the SIS 'offered to investigate the implication of cancellation of contracts with companies if necessary'.[33] It is further supported by the fact that the SIS did not envisage a refusal of al-Habobi's visa in June 1988 as jeopardizing its source.[34] Moreover, the issue of source protection does not figure in the FCO's consideration of the export of that first batch of machine tools.[35] In a memorandum dated 26 September 1989, the Head of OT2/3 said that source protection had been only one factor for the DTI even in the 1987/88 decision.[36] SIS witnesses told the Inquiry that Whitehall had, in the words of one of them, 'latched on to source protection to support a decision to preserve the trading relationship'.[37]

The claim that officials either failed to receive, or could not retrieve, the appropriate intelligence about the end use of machine tools is equally invalid. Everyone involved in the initial decision knew that the lathes were to be used to manufacture artillery and rocket shells. If, subsequently, no other intelligence had been received, the military end use of the machine tools could have been inferred from their going to the same destinations as before. As an internal FCO memorandum in June 1988 noted, 'We have been alert for some time to the nature of Nassr and Hutteen establishments and their connections to Matrix Churchill.'[38] As this knowledge was established in the minds of officials, so it also was in the minds of Ministers: 'We know,' wrote the Minister of State at the FCO in 1989, that 'high technology machine tools have been shipped to the major Iraqi munitions establishments.'[39]

Where there was specific intelligence about the end use of machine tools in Iraq's missile programme in 1989, officials cannot plausibly claim to have forgotten about it when writing submissions for their Ministers because of the *number* of reports, their *circulation* shortly before submissions were written, and the fact that the end use of the machine tools in a large ballistic missile project had been the subject of inter-departmental *discussion*. In any case, by 1989 at least, everyone in the decision-making process knew of Iraq's ambitions to be a state possessed of weapons of mass destruction and long-range ballistic missiles. Sir Richard Scott blames officials for not reflecting intelligence in the case of Project 1728 in submissions to Ministers. As a result of their omission, Ministers took the decision to allow the export 'on a false footing'.[40]

One has to search for reasons why intelligence was not reflected in submissions to Ministers. First, some officials argued that their Ministers would have found intelligence matters beyond their competence. For example, a civil servant from the MOD argued that the Minister was not told about the DIS's dispute with the SIS, in 1988, about allowing machine tools because to do so 'would mean the Minister having feel and expertise in the intelligence field to weigh one thing against another'.[41] The Scott Report was critical of this apparent failure, as well as the omission in telling the Minister about the collateral received for the original intelligence on the end use of machine tools.[42] However, Lord Trefgarne, then MinDP, asserted that he was 'entirely content' with the submission he did receive; and the civil servant concerned argued that the Minister would have known of the views of the DIS from the MODWG, which had DIS representation and input, refusing the machine tools applications on the grounds that they enhanced Iraqi capabilities.[43] The general proposition that intelligence reports were beyond the competence of Ministers is contradicted by the fact that Ministers in all three Departments did, in the ordinary course of events, see some intelligence reports.[44]

Second, officials, as well as Ministers, argued that intelligence was not necessarily a commodity to be trusted. However, this is contradicted by the readiness with which the Foreign Office accepted the SIS assessment of October 1989. At that point, its attitude towards exports of technology to Iraq had hardened, it was firmly opposed to allowing licences for the machine tools[45] and had known of the precise intelligence. The Scott Report explained the FCO's capitulation as stemming from Alan Clark's appointment as MinDP: this meant that the Foreign Office Minister now faced two protagonists for exports.[46] However, the FCO had had a perfect record in winning inter-departmental disputes about ELAs under the Guidelines, and yet, at the Ministerial meeting to decide upon the Matrix Churchill applications, the Minister of State conceded, instead of taking the issue up the Ministerial ladder as the FCO had successfully done in the past.

Ministers' claims that they would have acted differently had they seen intelligence relating to the machine tools' use in Iraq's missile programme are undermined, not only by the availability of intelligence for the fourth decision in July 1990, but also by their reactions to information about end use in other decisions. Sir Richard does not take into account this sort of circumstantial evidence which points to Ministerial responsibility rather than system failure. He judges only on the basis of the existence of intelligence reports on the one hand, and the absence of their substance in submissions to Ministers on the other.

The Minister of State at the FCO overruled the objections of his Assistant Under-Secretary of State, Middle East, to the export of machine tools in February 1989. Those objections, based upon a possible contribution to Iraq's nuclear weapons programme, were dismissed by the Minister of State, contemplating a visit to Iraq, with the comment that 'Screwdrivers are also required to make hydrogen bombs.'[47] Moreover, in September 1989, the Science, Energy and Nuclear Department of the FCO wrote a Minute, copied to the Minister of State about the machine tools export licence applications. The Department recommended refusal on the grounds that Matrix Churchill was 'part of the Iraqi procurement network' and 'the acquisition of advanced technology lathes could aid either a missile or a nuclear weapons programme'.[48] Here, then, is evidence of a Minister receiving two separate warnings about the possible end use of machine tools in the production of weapons of mass destruction and choosing to ignore them. In the second case, the Minister did not use the information from his officials in the subsequent meeting with his counterparts to decide the issue and instead capitulated.[49]

Ministers' claims to have acted differently are undermined by other examples. The MinDP was outraged to find the DIS, by its study of Iraqi procurement activity, 'interfering with a trade surplus'.[50] It was the Minister for Trade who encouraged machine tools manufacturers to fill in their export licence applications in a way which avoided stating a military end use. As the DTI's record of the meeting between the Minister and the Machine Tools Trade Association had it:

Choosing his words carefully and noting that the Iraqis would be using the current orders for general engineering purposes, Mr Clark stressed it was important for UK companies to agree a specification with the customer, in advance, which highlighted the peaceful (i.e. non-military use) to which the machine tools would be put.[51]

All present knew that 'the current orders' were to be used for a military purpose. The point of the meeting was the suspension of licences following the initial intelligence report. The MTTA were at no point asked for assurances that exported machine tools would not be used for military purposes. All subsequent ELAs for machine tools stated 'the precise purpose to which the goods will be put' as 'general engineering'.

A former Chief of Defence Intelligence has publicly stated that the assertion that officials did not pass on intelligence to their Ministers 'does not have the ring of truth'.[52] It remains to be seen why officials passed intelligence to Ministers in some cases but not in others. One is thus left with the conclusion that there was a general tacit understanding between officials and Ministers not to put in writing anything which could violently contradict the wisdom of the decision the Department wished to make. It should also be noted that it was officials who, in the main, opposed the grant of licences – the DIS, the MODWG, the IDC and senior officials in the FCO[53] – and Ministers who approved them.

The final batch of machine tools was licensed in July 1990 at a Ministerial meeting, chaired by the Foreign Secretary. From December 1989, the Foreign Office Minister had been refusing to approve machine tools for Iraq. At the official level, the FCO demurred on the basis that the Iraqi missile launch would make MTCR considerations much more significant and FCO Ministers would be less inclined to approve applications which might be for arms manufacture, even though they had approved them in the past.[54] MOD officials also opposed the machine tools licences but the Minister wanted to approve them. As the Departmental deadlock continued throughout the year, with Ministers from the DTI and the MOD opposing the FCO's position, not only did more ELAs for machine tools for Iraq become caught up in the delay but, as the delay continued, machine tools were in turn swept up in the review of the Guidelines which began in June 1990.

Some of the Matrix Churchill machine tools, as clearly stated on the export licence applications, were destined for Project K1000 at Nassr. The SIS asked Paul Henderson about the Project in April 1990, relaying to Henderson that the SIS believed it to be missile-related and suspected a British connection, i.e. Matrix Churchill.[55] There is no evidence that the DTI sought information from the SIS about Project K1000, although its very title would perhaps have prompted questioning. In fact, in December 1990, K1000 was found to be a nuclear weapons project.[56] However, to accompany the papers for the July meeting of senior Ministers, was a Joint Intelligence Committee report which noted that Matrix Churchill was 'also making parts for Iraq Project K1000, a possible weapons system'.[57] The JIC report was given to each Minister attending the meeting. Moreover, in the report was an overview of Iraq's intentions with respect to military procurement:

> Saddam's aim is to diversify Iraq's industrial base and achieve greater self-sufficiency in the production of strategic weapons. Industrial development has been and still is largely driven by military imperatives . . . There is increasing evidence of Iraq's ingenious and well-resourced procurement effort for military purposes directed to obtaining technology and specialised components for current weapons programmes.[58]

What is significant here is that, as the Scott Report noted, 'no one person was familiar with all the accumulated intelligence until, in June 1990, the JIC made the assessment'.[59] Ministers had the *complete* intelligence picture and yet decided to allow the export of machine tools.

The so-called 'Iraq Note', the briefing paper for the meeting, drew attention to the same data as had the JIC assessment:

> Iraq's policy is to supplement the resources of her indigenous defence industries and achieve the technology transfer she seeks through imports from the developed world, East as well as West, if necessary by clandestine means. Specific Iraqi import transactions identified in recent months, successful or attempted, include components for the supergun, capacitors, transputers . . . She is moving quickly to establish her own missile development production capability.[60]

This can only have served not as a sudden revelation but as a somewhat unnecessary reminder to Ministers of a matter that had for the past two years occupied the press and Parliament. Ministers had publicly acknowledged Iraq's procurement strategy in Parliament,[61] and Matrix Churchill, as those present at the meeting knew, was being investigated by Customs. The meeting, as the minutes record, was able to draw the lesson that:

> The acquisition by Iraqi Government interests of Matrix Churchill illustrated how industrial firms in the Western world could be used to reinforce the arms procurement networks of third world countries.[62]

Yet, the same meeting revised the Guidelines on defence sales to Iran and Iraq. The revision was to be held over because of Iraq's dispute with Kuwait. However, the meeting decided, with immediate effect, that the relaxation in the COCOM regime would apply equally to Iran and Iraq, thereby releasing from regulation the very dual-use goods, including machine tools, that Iraq sought to procure.

It was policy, not poor implementation of controls, that permitted such exports to Iraq. Each decision was made by Ministers and there is evidence that the Prime Minister was involved in at least one of the judgements about the export of machine tools. In a submission to the MinDP about an ELA for Matrix Churchill machine tools in December 1988, the Chair of the MODWG, Mr Barrett, referred to the Prime Minister's involvement. Lady Thatcher denied having been so, but the words of the submission are quite unambiguous. Of her involvement in the 1987/88 decision, it said 'The Prime Minister agreed that, in order to protect an intelligence source, the licences granted should not be revoked.'

In the same document, Barrett asserted that, because of attempts by al-Habobi to procure components for Iraqi nuclear weapons programmes, 'The case needs to go back to the Prime Minister before we could recommend approving the current applications.'[63] Asked about this at the Scott Inquiry, Barrett responded that he must have been mistaken, but added:

> it was not my practice to refer to the Prime Minister without good reason . . . it was not a typing error . . . I certainly didn't make it up . . . None of the copy addressees . . . raised it with me.[64]

The decision taken in July 1990 was executed after it was ratified by the Prime Minister who took a close interest in Iraq's procurement. The activity displayed by officials in passing machine tools decisions to Ministers can be contrasted with what one is asked to believe was the narrative in the supergun affair. In the Chair of the MODWG's reference to the Prime Minister one can again see the sensitivity of officials to the policy concerns of the centre. That the centre holds sway is demonstrated in the MODWG's recommendation to the Minister *vis à vis* Iraq's military procurement:

> From the point of view of both military and security concerns, the MODWG would like to see a tightening up of the Guidelines on the release to Iraq of items which could be associated with the setting up

of an arms industry. *However the MODWG recognises the existence of political and economic factors and that at the moment, they are not allowed to be as strict as they would wish.*[65] (emphasis added).

# 9

# The Iraqi procurement network and British motives

## Conventional arms and weapons of mass destruction

The UK saw no obvious disadvantages in selling arms to Iraq. Indeed, allowing the export of the initial parts of the supergun might have been based on a belief of the conventional nature of the weapon. In the case of dual-use goods which contributed to Iraq's conventional military industry, the British Government also perceived no negative consequences. The Head of the DESS commented: 'In terms of a build-up of indigenous capability, that is not recognised by anyone as a bad thing *per se*.'[1] And the Minister of Defence Procurement: 'I would not have raised any *prima facie* objection to our helping Iraq build up a conventional arms industry.'[2]

The British contribution to Iraq's indigenous defence base might have served much the same sort of foreign policy goals as the UK's overt and covert supplies of arms. As Lt Col. Glazebrook commented: '(The) USSR was, at that time, our enemy . . . and anything that could reduce the linkage between Iraq and the USSR was considered a good thing.'[3]

Furthermore, in the context of the Iran–Iraq War it was, as Treasury Counsel put it 'not a wholly fanciful argument' that the Government allowed the export of the supergun tubes 'for political reasons'.[4] Similarly, while the war persisted, the UK might have perceived an advantage in supporting Iraqi arms production in its fending off Iran. Certainly, Alan Clark put it in those terms to Sir Hal Miller: 'the Iraqis were the "goodies" in the Iran–Iraq War'.[5] Machine tools in 1987 and 1988 went to Iraqi artillery production and, as Cordesman and Wagner have asserted, 'Iraqi defensive tactics emphasised massive application of firepower.'[6] In any case, as the DIS argued, artillery is 'decisive' in battle.[7] The supply of machine tools in 1987 and 1988 coheres with Alan Clark's assertion that the UK, concerned at Iranian preponderance from 1987, increased its support for Iraq. However, the evidence remains ambiguous, not least because Britain knowingly supplied machine tools to Iran for military end uses.

It was only the DIS and the MODWG who made the arguments that, in contributing to Iraq's indigenous military industry, the UK was both assisting the creation of a competitor in the defence market and undermining the British ability to exercise influence by withholding arms supplies. A further objection voiced by the two bodies was on the basis of possible diversion to the Soviet Union.[8] However, such arguments, concerned as they were with the long term, were swept away by more immediate preoccupations.

Where the UK's short- to medium-term security was not at stake, the impetus was to export. Constraints, however, did exist, not only in the guise of national security, but also as other foreign policy considerations. International control regimes reflect the UK's insecurities both in seeking to limit proliferation in WMD and ballistic missiles and, before the end of the Cold War, in seeking to keep a technological, and therefore military, lead over the Eastern bloc. However, there was a tension between that security interest and the need to trade. It was, and is, the pressure from allies, particularly the US, which ensured compliance with regimes where interests conflicted. This is one such foreign policy consideration. Another is the UK's foreign policy interest with a recipient state's foes or neighbours. The operation of this concern was noted in the discussion of British arms sales to Iran and Iraq. It is possible that such considerations influenced events in the supergun affair.

Parts for Baby Babylon, the 350mm gun, were exported, sufficient to allow four test firings of the weapon. However, the supergun was not completed because of the seizure of the final consignment of eight tubes and the explosion at PRB. The 350mm gun had a range of 390 km and a payload of 15 kg; the 1000mm gun, 700 km and 390 kg. While MOD experts concluded that 'a proliferator state might have difficulty' in manufacturing a nuclear warhead small enough for the larger gun, 'Iraq could have developed chemical and/or biological warheads for the supergun project'.[9] Moreover, as the intelligence report of the supergun (Appendix B) noted:

the Iraqis would appear to have adopted a proper systematic approach (words redacted) which could lead to a successful outcome given continued access to the necessary resources and West European technology.

It is difficult to see the seizure of tubes at Teesport and the explosion at PRB as two fortuitous and independent events. The words in the intelligence report above, as suggestive of close monitoring, seem to indicate that they were not. Thus, it can be argued that the UK, while content to see Iraq gain conventional capabilities, was prepared to act to prevent the acquisition of non-conventional ones. Whether or not the UK saw a long-term threat to its own national security is unclear.[10] However, states in the Middle East, had they known of the supergun's existence, would have been immediately threatened by it. Pressure on the UK from Israel, or the US on Israel's behalf, to act before the super-gun became operational cannot be discounted. The seizure of the tubes came one week after Saddam Hussein threatened Israel with being eaten by fire – at least that was the American interpretation of Hus-sein's warning of chemical retaliation for an Israeli nuclear attack.[11] Nor can Saudi Arabian influence be excluded. The kingdom knew of the gun because Iraq had had to negotiate to use part of Saudi territory as a firing range.[12] While the exigencies of the Iran–Iraq War might have persuaded Saudi Arabia to co-operate, post-war, a confident, well-armed Iraq with a supergun could have been more than the kingdom could bear.

Foreign policy, interpreted as the security of both the UK and its friends, did operate as something of a constraint upon the dual-use technology supplied to Iraq. This concern is expressed in the forma-tion of the Working Group on Iraqi Procurement. There is nothing to suggest that the UK did not, as a rule, respect the international control agreements. It had a national security interest vested in them, and an interest in maintaining its international reputation in being seen to uphold them. For example, the FCO would not countenance the export of hydrogen fluoride to Iraq even though it was 'such a widely traded chemical' because it was on the Australia Group Warning List and 'partners could be critical if they became aware of the supply'.[13]

Again, the FCO and the MOD advised against the supply of hydro-gen fluoride to Egypt after Israeli intelligence about diversion to Iraq because of 'the leading role the UK has consistently taken in interna-tional efforts to curb CW proliferation'.[14] The Minister of State advised that, *vis à vis* Iraq's application for a grant for the Learfan fac-tory, the Government faced 'a scandal based on reports that, when we keep a high moral tone about the Germans and others, we have Iraqis based in Northern Ireland in an MTCR technology'.[15]

The FCO was also more careful about domestic public opinion. Its objections to the export of machine tools began to crystallize at about the same time that press coverage of both Iraq's procurement network and its appalling human rights violations became widespread and frequent.

Nevertheless, the UK's interest in preventing proliferation is not a superficial one. It has sought to support the international control regimes both with statute and with guidelines. Mention has been made of both the NBC Guidelines, which introduced quota systems for all countries and outlawed the sale of defensive NBC equipment to Iran and Iraq, and the WMD 'Countries of Concern' List. The latter, although following a US and German lead, was formulated in part to persuade other members of the Australia Group to do likewise.[16] However, it was where international regimes did not exist, or where there was ambiguity, either in the regimes themselves or in the possibility of diversion, that other more immediate interests came into play.

## The absence of control regimes

Officials blamed the failure to check Iraqi procurement from the UK upon the Cold War: the focus was upon COCOM instead of upon proliferation.[17] This explanation will not do. It was the interest of trade which overwhelmed restraint when no clear, immediate, countervailing interest existed. This must be seen in the international context of competitive states, other suppliers: it is the 'prisoner's dilemma' in other clothes. For example, Lowther contended that, given the absence of any international regime or statutory control, the UK was content to see Iraq buy the Learfan factory in Northern Ireland. Lowther quoted a letter sent by the Foreign Office to Kevin McNamara MP in which it argued that the *United States,* as a member of the MTCR, was committed to the prevention of the proliferation of ballistic missiles. Lowther quoted 'sources' who explained the reference to the US as reflecting the fact that it was American pressure which forced the UK to act. The SIS was content to see Iraqi–SRC ownership since they had a contact in Dr Bull's company and 'felt it better to have Iraq involved in a project it could monitor than in one it could not'.[18] If Lowther is right, it was this ambiguity of the purchase of Learfan – uncontrolled by statute or by international agreement – which allowed other factors, other than the non-proliferation effort, into the decision-making process.

The export of hydrogen fluoride, referred to above, *was* permitted, in spite of the strong strategic and political interests in not doing so, because of more pressing, countervailing interests. The Minister of State was about to visit Egypt, and a refusal would have caused 'an unfavourable atmosphere'.[19] Potential diversion created another ambiguity, a space into which other considerations could flow.

The DTI argued that:

> policy in the missile (or) nuclear control context should be aimed at preventing the export of critical components, not at general industrial equipment, unless there was evidence of direct involvement.[20]

It was the inherent ambiguity of machine tools and the fact that they were not controlled for the purposes of non-proliferation that permitted the trade interest to prevail. Even where intelligence existed, it could be disregarded because intelligence, by its very nature, does not represent evidence of a direct and certain quality. For example, Ministers and officials chose to believe that, following the ceasefire, Iraq was switching from military to civil production.[21] The then MinDP said that this was based upon 'common sense'.[22]

## Iraqi procurement and British trade

One can divide the UK's motives in exporting dual-use goods to Iraq into two aspects: positive and negative. Sales of dual-use equipment to Iraq were made for positive reasons to do with the trading position of the companies concerned and the general desire to improve the UK's share of the Iraqi market. Such motives were prevalent where there was no concern about the use to which exports would be put – for example, in contributing to Iraq's conventional military industry. Negative pressures encompassed anxieties about Iraqi debt repayments and direct coercion from Iraq in the shape of threats to, or acts which disrupted, civil trade. These negative influences occurred, and were effective, where the UK was more reluctant to supply certain dual-use technologies.

In the case of machine tools, there was a desire to assist the manufacturers. A senior FCO official had the following exchange with Sir Richard Scott:

> A. I thought the commercial and employment aspects of this particular sale were important.
> Q. More important than worrying about whether or not there was going to be a significant enhancement to the Iraqis' munitions making capability?
> A. Yes.[23]

Britain sold more machine tools to Iraq between 1987 and 1989 than France, Italy and the US combined, although Germany outsold Britain, as did Switzerland, with the exception of 1988.[24]

An examination of the top ten UK exports to Iraq reveals that it was industrial and scientific equipment which represented the bulk of trade. In 1990, such goods accounted for £223,954,000 of a total of £293,393,000.[25] Iraq was an important market – by 1989, the UK's third largest Arab market[26] – and was courted by Ministers accordingly. Except for 1982, a Minister visited Iraq in each year of the 1980s with the FCO aiming for a visit by the Foreign Secretary in 1990.[27] Iraq, with the loss of Iran, was 'the big prize',[28] especially when it wanted the very British manufactures that were most in need of support. Alan Clark is worth quoting at length on this point. He, in turn, is quoting a Conservative whip on the supergun:

> 'We should be making them, and selling them to everyone . . . Good God . . . All this stuff about a decline in our manufacturing capability, but they had to come here to get the barrels made, didn't they? We should put it in a Trade Fair.' Splendid fellow.[29]

The light-hearted tone of the above should not obscure a genuine sentiment. Walter Somers' and Sheffield Forgemasters' contracts with Iraq were worth £1.9 million and £8.4 million, respectively.[30]

The source of Iraq's purchasing power, its oil, must also be borne in mind. The UK imported no oil or oil products from Iraq in 1987 and 1988. In 1989 it imported oil and associated products to the value of £18,080,000. In 1990, the figure was £85,489,000.[31] Although this represented roughly a quarter of British oil imports from Saudi Arabia and a third from Iran, the price at which Iraq sold oil to the UK was lower than from any other Middle Eastern source. Oil exports from Iraq to the US recorded similar sharp rises at favourable prices.[32] In 1989, the first major post-war energy sector contract went to a British company, and the Oil Exploration Company of Iraq and British Gas signed an agreement on increased technical co-operation.[33]

By the end of the Iran–Iraq War, Iraq had built up a significant debt with the UK, as it had to other countries. The CIA estimated that Iraq was some $35 billion in debt by 1989, whereas its per capita income was only $1,940.[34] In essence, repayment was problematical, not least because Iraq's development programmes remained equally ambitious and reliant upon imports. The trade sought by industrialized states, its oil, together with its indebtedness, meant that Iraq was able to exercise influence over its creditors, both for rescheduling and for new credits. As the CIA's Directorate of Intelligence put it in 1990: 'Baghdad continues to repay its debt on a selective basis, giving priority to suppliers of key goods or new credits as well as to important political allies.'[35]

More importantly, Iraq was able to use debt as a lever in its procurement strategy. As a British diplomat said of France: '(They) are perhaps obliged to go on supplying things against their better interests in order to make sure that they get paid.'[36] However, while Iraq's debt to France was greater than that to the UK – in 1990 $6 billion – the same leverage occurred.[37] If the US was persuaded to continue its Commodity Credits for fear of inadequate servicing of the $2 billion debt,[38] it can be assumed that a much less wealthy country such as the UK was similarly, if not more, influenced. While the Treasury had argued, in March 1990, that the Government 'should cut (its) losses now and close down the credit line', it had been overruled by the DTI, the MOD and the FCO which feared that Iraq 'would hit back hard'.[39]

The Secretary of State for Trade and Industry reported Iraqi indebtedness in June 1990 as £2 billion of risk and £140 million in outstanding payments.[40] In his letter to the Prime Minister, which triggered the revision of the Guidelines on defence sales to Iran and Iraq, the Secretary of State argued:

> On the one hand we need to minimise our involvement in the Iraqi procurement programme. But we also need to bear in mind the implication of export controls on our exports to Iraq and on ECGD's large exposure on that market ... While Iraq has hitherto treated the UK as a preferred creditor the present high level of arrears reflects the cessation of payments during the last two months which was evidently linked with the current political coolness ... Consequences of a systematic Iraqi default would clearly be extremely serious for ECGD and would have implications for the Public Sector Borrowing Requirement.[41]

Iraq applied general and specific pressure on the UK to continue exports of dual-use technology. The Iraqi Ambassador raised Matrix Churchill's ELAs with the FCO in January 1989, a few weeks before the Minister of State was to visit Iraq. A few days before Waldegrave was to leave, he overruled his senior officials' objections to the machine tools exports.[42] The Minister of State had also recognized the general point that Iraq's continuation as a large market for British goods would depend upon the health of the political relationship. In turn, that relationship depended upon continuing exports of defence-related dual-use goods. Waldegrave argued:

> I doubt if there is any future market of such a scale anywhere where the UK is potentially so well placed if we play our diplomatic hand correctly, nor can I think of any major market where the importance of diplomacy is so great on our commercial position.[43]

The sensitivity of Iraq to each export licensing decision of significance to it is evident in the content and language of the DTI's submission on the second batch of machine tools in 1989. A refusal would be seen as 'provocative' by Iraq.[44]

It was increasing pressure exerted by Iraq following the seizure of parts of the supergun which led, in July 1990, to the relaxation of controls. The Iraqi Ministry of Industry and Military Manufacture, which accounted for about 60 per cent of industrial procurement, had announced that trade with the UK was under review – in effect, an embargo on new business. In spite of British assurances from both the Ambassador and the Secretary of State for Trade and Industry that Britain was neither interfering with civil trade nor co-ordinating an anti-Iraq campaign, the embargo continued.[45] The Iraqi Government stepped up the level of anti-British rhetoric, sponsored demonstrations by its citizens and persuaded other Arab states that Britain, in its Customs and Excise investigations, was conducting a vendetta against Iraq. Arab states joined in Iraqi condemnation of the UK. Iraqi contracts worth about £170 million were estimated as lost to British business, which in turn expressed displeasure to the Government.[46] The DTI was anxious that the Iraqis should not find out about Customs' investigations of Matrix Churchill which began in June 1990.[47]

The decision to relax controls on dual-use goods to Iraq in the wake of the accumulating evidence of Iraq's procurement of weapons of mass destruction and ballistic missiles indicates the conflicting interests which surround defence-related sales and which particular interests are strongest at the point of decision. Here, as in most cases, the strongest motives are those of the short-term, not the long-term. If appeasement – the most despised of all post-war policies – can be seen in this decision, it is because short-term expediency is the essence of appeasement.

## The international context

However, export licensing decisions regarding dual-use technology for Iraq must be seen in context. The thread running through all official and Ministerial argument about ELAs for dual-use exports to Iraq was that refusal was pointless if other countries would supply. France was Iraq's major Western supplier of arms, and also made a significant contribution to Iraqi nuclear development. The 'prime suppliers' of Iraq's chemical, as well as nuclear, programmes were the West Germans.[48] Switzerland was the main supplier of precision machine tools and, like Italy, Belgium, Holland and Spain, supplied components for the super-gun as well as for other projects involving the development of weapons of mass destruction.

The US, the leader of the international non-proliferation regimes, approved, between 1985 and 1990, according to the House of Representatives Committee on Foreign Affairs, '771 licenses for the export to Iraq of $1.5 billion worth of biological agents and high-tech equipment with military application'. There were only thirty-nine refusals in this period. While the Committee noted US foreign policy concerns while Iraq was at war with Iran, it reported that exports continued until the invasion of Kuwait, despite severe problems in the Iraqi–American relationship. The Committee concluded that, 'The Export Control System did not break down. Saddam Hussein got the equipment that the State Department wanted him to have.'[49]

Senator Gonzalez, the Chair of the Committee on Banking, Finance and Urban Affairs, concluded, at the very least, that the CIA had 'contemporaneous knowledge' of BNL's funding of Iraq's procurement programmes.[50] The Atlanta District Court suspected an international conspiracy.

However, it is the element of competition, rather than conspiracy, which should be stressed. It was, as the DTI said, 'a buyer's market'.[51] Ministers were constantly reminded in making decisions on dual-use goods that there was 'no shortage of foreign competition'.[52] Such attitudes were neither confined to the DTI nor to officials. The Foreign Secretary, in allowing al-Habobi a visa, argued that, because of the extent of al-Habobi's business involvement in the UK, 'If he were excluded . . . we (would) lose valuable contracts in Iraq.'[53] In a buyer's market, abstinence must be its own reward. Restraint is only worthwhile if a Government has the certainty that others will do likewise: that is, international control regimes are essential. In the gaps between them, proliferators can operate: the supplier is more influenced than influencing. Even when there was suspicion that precision machine tools might be going to Iraq's nuclear programme, SEND was forced to concede that, as the lathes were not listed in either the Nuclear Suppliers' Guide or the MTCR, and only appeared in the Sensitive Nuclear Explosives Technology List of which only the UK, US and France possessed a full list, the concern of other Departments that other suppliers would fill the export was 'valid'.[54]

In seeking to persuade other countries not to supply certain equipment, a Government needs hard evidence, which is difficult to come by, if it is to persuade a competitor to forgo exports. The UK was suspicious of intelligence from its competitors.[55] Such intelligence sharing would, in any case, have been problematical given that, in the British case, and probably in those of other countries, it was obtained by participation in the very trade that was of concern – viz Matrix Churchill. The intelligence agencies have a vested interest in the continuation of 'grey' exports in order to maintain their sources, and decision-makers seem sympathetic to that position.

The other constituent of the international environment is the existence of other recipients. Iraq was unique only in having used its technology from the industrialized world so soon after acquiring it, in a war with the international community. It was not unique in seeking either an indigenous defence base or WMD and ballistic missiles. There is evidence to suggest that Iran was similarly supplied with dual-use goods by the UK, as well as by other European countries.

Britain's involvement in Iran's indigenous defence industry began in the days of the Shah when it was the primary supplier of a weapons plant at Isfahan, valued at $1,400 million.[56] By 1987, Iran was producing over $200 million worth of arms, mainly ammunition, mortars, multiple rocket launchers and small arms.[57] Britain had supplied Iran with machine tools for arms manufacturing. An SIS telegram dated 18 January 1988 noted: 'DTI had no objection for export of machine tools to (words redacted) although it was known that an armaments factory was the destination.'[58] This could be a reference to Nassr or Hutteen, but these names appear elsewhere in the text and are not redacted. The sentence quoted follows one which mentions West German exports to 'both sides'. At the MTTA meeting in January 1988, the Minister for Trade, in illustrating his point about the need to agree a 'peaceful' use, gave the example of Iranian imports of machine tools for agricultural machinery plants.[59]

Such exports have continued. A UK machine tools company, UBM, withdrew from a contract with Iran when it realized its machine tools would be programmed to rifle 155mm Bull-designed gun barrels. The export had DTI approval, despite the company expressing its fears to the Department and the machine tools' use being obvious.[60] Another machine tool manufacturer, Wickham, has exported lathes to Iran.[61]

Iran was acknowledged within Government as a proliferator state.[62] In 1989, the owner of a specialist electronics company wrote to the Prime Minister for advice about the supply of components for monitoring and controlling gases, including hydrogen cyanide. The Minister for Trade responded, saying that the Government 'had no reason to believe that these sensors would be used for a military or other applications which may be contrary to the UK's interests'. The owner's suspicion, communicated to Downing Street, was later confirmed in correspondence with the Iranian Defence Industries Organisation.[63]

In February 1990, *The Observer* reported that the US was applying pressure on the UK to prevent the supply to Iran, by a British company, of a phosphorus pentasulphide plant, worth $37 million.[64] The chemical was brought under British statutory control in August 1990 as a precursor for VG and VX gas: it had been on the Australia Group Warning List from May 1989. The Minister of State at the FCO mentioned it to the Minister for Trade in a letter dated 27 April 1989 dealing with other matters. He said: 'I have just received your separate letter on the Phosphorous Pentasulphide plant for Iran, which we shall pursue separately.'[65] The comment is suggestive of a disposition to consider the export. Furthermore, it was not until almost a year later that *The Observer* reported American pressure to stop the export. This suggests that, at the very least, the application was not turned down immediately, if at all. It is interesting to note that it was considered, apparently solely, at Ministerial level.

Only in 1993 did the UK review its guidelines on exports to Iran, preventing the supply of Military and Atomic Energy List goods. Industrial List equipment can still be exported, provided the end use is non-military.[66]

## Summary

There is no reason to suppose, therefore, that Iraq was *sui generis* in benefiting from Western, or Eastern, technology in its conventional and non-conventional procurement effort. The fact that Iraq was not a unique case increased the pressure to export. Nor is there any reason to believe that, in a competitive international environment, other countries, even the US, were not motivated by the same forces which drove the UK to supply Iraq with dual-use technology. There was no failure in the processes of control. Instead, there was a series of political decisions about where the balance of interests lay. Since democratic governments come and go on the basis of perceived economic performance, it is not surprising that short-term goals of trade overwhelmed longer-term economic and strategic potentialities.

The same forces were at work in the export of dual-use technologies as in the export of conventional arms. The motive to export represented a mix of overlapping elements of influence and trade. Whereas in the case of exports of conventional arms, the decision-maker placed emphasis, all other things being equal, upon the export as commodity rather than as arms, in the case of dual-use goods it was so much easier to focus the perception upon the civil, rather than the military, uses to which exports might be put. This was only the case, however, where there was any ambiguity. But, it was this ambiguity which created the vacuum into which immediate wants – the desire for influence or for trade – entered. For the UK, defence export decisions represented the outcome of a collision of national interests.

# Conclusion

Neither the bureaucratic politics nor the organizational process model of state behaviour is useful in explaining British decision-making about arms exports. Rather, one should look to political and economic circumstances. It is this context – in essence what the centre perceives is necessary or desirable – which gives power to some in the decision-making process over others. The inter-departmental decision is not a chaotic contest for power between competing groups. Policy was made by the centre; and the process operated to bring together the UK's competing interests in decisions about weapons exports. Departments manoeuvred *within* the framework of policy, not without. Ultimately, shared values were more important than divergent ones: all Departments shared the perspective that arms exports policy should serve Britain's economic and physical security. The contest between Departments reflected competing British interests in the varying circumstances in which judgements were made.

In the 1980s, the UK operated an official policy of restraint with regard to arms sales to Iran and Iraq and an unofficial one of supply. The official Guidelines were aimed not at the progress of the Iran–Iraq War but at the maintenance of relationships with the belligerents – hence an apparent policy of evenhandedness. They represented a framework for the examination of the balance of British interests depending upon the exigencies of the time. The Guidelines were a procedural means to a policy end.

If analysts have overlooked the complexity of the UK's motives in supplying arms, it is because British interests are rarely revealed in its policy declarations. The Guidelines on defence sales to Iran and Iraq are an exemplar in this respect. Ostensibly a policy for bringing the Iran–Iraq War to its earliest conclusion, and for fulfilling the requirements of neutrality, the Guidelines deliberately disguised the UK's real interests. Like most arms export decisions, individual judgements about sales to Iran or Iraq represented a balance between competing interests, although foreign policy was pre-eminent. The main engine behind the Guidelines was a general desire to trade; and this acted as both a force for sales and for restraints, with pressure coming from all directions. Arms brought influence in that they provided access to the general market. However, the importance of that general market then became a tool of influence for the recipient.

The UK's participation in the illicit trade in arms with Iran and Iraq represented an extension of its official policy in serving the same interests. The Government's knowledge of widespread diversion of military goods by Saudi Arabia, Egypt and, in particular, Jordan to Iraq is clear from the abundance of available evidence. It is possible to establish the probability of the Government's knowledge of British participation in the European arms cartels either through the Swedish Customs investigation, the French investigation of Luchaire, or through the involvement of a publicly owned company, the Royal Ordnance Factory. More generally, the likelihood of the Government's knowledge of illicit British arms sales to Iran and Iraq is evidenced by both the general state of its understanding of trans-national conduits, false end user and shipping documentation, and the fact that the intelligence agencies would have had at least some knowledge of the trade. This was especially so in the case of Iran given the ease with which the activities of the Military Procurement Offices would have been monitored.

Moreover, established British companies knew that the price of government support for exports was openness. The frequency with which participants have claimed to have alerted the Government to contracts for Iran and Iraq cannot be discounted. Furthermore, as new evidence has periodically emerged, it has been the accounts by those exporters, rather than that of the Government, which have been verified.

Given the state of government knowledge on the one hand and, on the other, the fact that the system of reviewing ELAs for Iran and Iraq expressly excluded ELAs for other destinations which might be diverted, one has to conclude that illicit supplies represented policy and not failures in control mechanisms. The Government's motives in allowing the illicit supply of arms to Iran and Iraq were varied. Illicit sales would only have damaged British interests if Iran or Iraq, or Iraq's supporters, had perceived the Government's connivance in supplies to the other party. Such sales would have assisted the pursuit of both influence and trade, and participation in covert supply would have provided an important source of intelligence as British business people acted as informal agents. As the discussion of the official policy showed, the distinction between 'lethal' and 'non-lethal' defence goods was at best subjective and at worst meaningless. Claims that officials failed to keep Ministers properly informed must be judged in this context.

In turning a blind eye to Jordanian and Saudi arms export diversion, in particular, the evidence suggests a general concern to maintain and increase the defence trade with Arab countries. As for Iran, given the strong opposition by the US and Arab states, illicit sales can be seen as bridging the gap between British interests with important friends and trading partners and Iran's economic and political significance. Moreover, allowing arms to reach both Iraq and Iran meant that those markets were at least kept upon a 'care and maintenance' basis. There was, of course, an immediate opportunity cost in not supplying arms to the belligerents, especially when other countries continued to export.

The UK was not motivated in its arms export policy to affect the course of the Iran–Iraq War, although it is possible to establish an interest, at least until 1987, in sustaining the conflict while it remained contained. Three factors undermine the thesis that the UK sought, via its arms export policy, to prolong the Iran–Iraq War. First, given the number of suppliers, British arms could only have been locally and temporarily important. Second, given the fact that the war was one of attrition and marked by poor use of technology and tactics, the British could not have been certain of the effectiveness of its arms shipments. Third, what was supplied illicitly was necessarily limited by the Government's desire not to be seen as conniving at such exports.

It might be argued that Britain's arms supply represented a contribution to an international effort to prolong the conflict. This, however, is undermined, not only by the unpredictability of the course of the war, but also by the fact of the American effort from 1983 to end the War in order to protect Iraq and the Gulf from Iran. Soviet interests in both containing Iran and ending the war further mitigate against the international conspiracy theory. Rather, one should stress the *competition* between supplying states for shares in the political and economic marketplace which the Gulf's killing fields represented.

The same factors which drove the UK's sale of conventional defence goods drove the sale of dual-use equipment to Iraq and Iran. The most important of these factors were trade and political relationships. The strength of these motives was to be seen in the UK's inactivity when confronted with evidence of Iraq's procurement of WMD and ballistic missiles. The British contribution to Iraq's indigenous conventional arms industry was not seen by the Government as necessarily something to be avoided. Indeed, it might well have served the same foreign policy interests as the supply of overtly defence equipment. However, foreign policy, as the UK's security or that of its friends, intervened to prevent Iraq's operationalizing the supergun. Furthermore, there is little evidence to suggest that the Government did other than adhere to international control agreements, both for the sake of its own national security and for the sake of its international and domestic reputation. It was where there was ambiguity that other immediate interests overwhelmed the longer-term concerns in constraining sales of dual-use technology to Iraq's military projects. This must be seen in the context of a world of other suppliers.

The UK's supply of dual-use goods to Iraq can be seen as being driven by positive and negative forces. The former category included a desire to support particular companies and a more general aim to improve the UK's share of the Iraqi market. These positive motives operated where there was no concern about the purpose of dual-use goods. Negative pressures included concern about Iraq's manipulation of its debt repayments and its overt threats to Anglo-Iraqi trade. These pressures were successfully applied where the UK was unwilling to supply some dual-use technologies. In short, the UK operated a policy of appeasement towards Iraq.

Given the ambiguities intrinsic to dual-use goods and the fact that the harm to British strategic and economic interests was in the long term, it is unsurprising that such interests were overwhelmed by immediate wants. In such cases, as in those involving overtly military equipment, defence export decision-making was less about calculated policy and much more the product of a clash of national interests.

Three objections might be raised to the wider relevance of this case study. The first is the particular ideology, or world-view, of the British Government in the 1980s. As Bernard Porter asserted, 'The climate was "free marketism", and the foreign policy it gave rise to was a competitive internationalism'.[1] This ideological perspective, moreover, was laid over the cognitive dissonance[2] of Britain's relative decline. However, while arms sales might be seen as part and parcel of a new proactivism in British foreign policy, it must be assumed that if Britain has found arms exports to be a useful tool of foreign policy, other states have made the same discovery. Furthermore, Britain is far from alone in carrying the historical legacy of empire. Kolodziej has noted a similar influence in respect of France.[3]

The second objection to Britain's usefulness as a case study is the weakness of the UK economy. The UK, as a former Foreign Office Minister put it, 'exports or dies'.[4] While it is true that exporting is a special preoccupation in the economic discourse of Britain, it can hardly be said to be insignificant to other industrialized countries. Indeed, the commitment to export can be seen in the subsidized competition for markets which the UK faces from Japan, Germany, France and Italy. In any case, trade is the basis of state power and while some countries might be able to exercise more restraint of trade than others, Britain can at least serve as an exemplar for most arms suppliers in the inter-related quest for defence and civil markets.

Finally, it might be argued that Iran and Iraq should be seen as *sui generis* as recipients of arms and technology. First, it might be argued, they were able to exert pressure because of their importance, as markets, as oil producers, and as powerful states in an internationally significant region. Second, it could be said that foreign policy was so predominant in defence exports policy because of the sensitivity of the situation in the 1980s, that is, the fact of the Iran–Iraq War.

With respect to the first argument, Iran and Iraq each represented about half of 1 per cent of total British exports in the 1980s: *inter alia* Oman, Turkey, Greece and Nigeria represented similar percentages. Iran and Iraq were not unique in terms of the export markets they represented. The special significance of their oil reserves is negated by the fact that it was their ability to use not oil exports but civil trade as a lever in ensuring that the supply of defence and dual-use technology was continued by the UK. While the UK imported oil from Iraq and Iran, the quantities were small. However, one cannot rule out the pursuit of influence with both states for the sake of the security of Middle Eastern oil as a whole. While Iraq, and especially Iran, were important regional actors, similar states can be found in other regions and it is arguably the ability of such states to gain arms and arms-making technology which is most significant in the international system. The second argument can be countered by pointing to the potential for conflict around the world: neighbouring countries will always be sensitive to each other's arms imports. Thus, foreign policy will continue to be paramount in defence exports policy.

This book remains a case study, and is thus limited. However, it is significant in that, through utter serendipity, it represents one of the few in-depth studies of the *motives* of a major arms supplier and one that has had the benefit of access to primary sources from the decision-making process. This case study, to reiterate, indicates that five propositions require further research.

First, arms exports, for secondary suppliers, are not primarily driven by the structure of the defence manufacturing base but from a more general desire to export. Second, *all* suppliers are motivated by overlapping reasons of foreign policy – influence and trade – in exporting arms and dual-use technology. Third, it is not true to say that only the superpowers could refuse arms transfers exclusively for political reasons.[5] Fourth, civil and military trade are interrelated in that the willingness to supply defence goods becomes the price of access to the wider civil market: reverse influence continues to obtain. Fifth, outside of international control regimes, arms export policy does not exist: decisions are reactive, involving conflicting interests.

Further case studies are both necessary to support the conclusions of this book, but are likely to be difficult to conduct given the lack of information. At the very least, the focus upon the exigencies of the defence manufacturing base as the cause of arms exporting for second-tier suppliers needs to be changed, and arms exports located more widely in international relations.

In terms of control of the arms trade, future research should be aimed at the interests in, and the mechanics of, strengthening international agreements that confront proliferation. Given the pressures which potential recipients can exert over suppliers of arms and dual-use technologies on the one hand, and the complex interests for suppliers in exporting such goods on the other, that research represents something of a challenge. As for the conventional arms trade, this study suggests that the picture is not as stark as has been thought. The arms trade is not de-politicized; it is not free of restrictions. Control might be exercised for narrow national interests rather than for ethical considerations, but domestic and international pressures for restraint can be brought to bear upon the suppliers of arms.

# Notes

## Introduction

1   *Official Report*, 29 October 1985, Vol. 84, col. 454.
2   John Goulden CMG, Assistant Under-Secretary of State at the Foreign and Common-
    wealth Office, Trade and Industry Committee, *Exports to Iraq*, Minutes of Evidence,
    28 January 1992, Session 1991–92, House of Commons Papers 86–viii, p. 281; and
    *Defence Sales to Iran and Iraq: Proposed Revision*, 20 December 1988, (000111)*.
    (* Numbered documents refer to those released to Paul Henderson and passed to the
    author.)
3   Trade and Industry Committee, *Second Report, Exports to Iraq: Project Babylon and
    Long-Range Guns*, Session 1991–92, House of Commons Papers 86.
4   The Rt Hon. Sir Richard Scott, the Vice-Chancellor, *Report of the Inquiry into the
    Export of Defence Equipment and Dual-Use Goods to Iraq and Related Prosecutions*,
    House of Commons Papers 115, 1996. (Hereafter referred to as *Report*.)
5   *Ibid*, D8.16, p. 816.
6   *Ibid*, K8.1, p. 1799.
7   *Ibid*, D8.16, p. 816.
8   *Ibid*, G13.125, pp. 1395–96.
9   *Ibid*, E2.62, p. 851.
10  *Ibid*, D6.169, p. 643.
11  *Ibid*, D3.34, pp. 386–7.
12  *Ibid*, F4.80, p. 1095.
13  *Ibid*, K8.15, pp. 1805–06.
14  House of Lords Select Committee on Overseas Trade, *Report*, Session 1984–85, House
    of Commons Papers 238, pp. 47–8, quoted in Andrew Gamble, *Britain in Decline: Eco-
    nomic Policy, Political Strategy and the British State*, Fourth Edition, London, St.
    Martin's Press, 1994, pp. 208–9.
15  *Inquiry into Exports of Defence Equipment and Dual-Use Goods to Iraq*, Hearings in
    the presence of the Rt Hon. Lord Justice Scott, Record of the MOD Working Group
    meeting, 13 December 1984, quoted in Day 4, p. 29.
16  *Ibid*, Evidence of Stephen Day, Day 1, p. 194.
17  *Ibid*, Evidence of Sir Richard Luce, Day 1, p. 4.
18  Paul Henderson, *The Unlikely Spy*, London, Bloomsbury, 1993, p. 257.
19  See: D Sanders, 'Government Popularity and the Next Election', *Political Quarterly*,
    Vol. 62, 1991, pp. 235–61.
20  *Report, op cit*, K2, K3, K7 and K8.
21  National Audit Office, Report of the Comptroller and Auditor General, *Ministry of
    Defence: Support for Defence Exports*, House of Commons, Cmd 303, 1989, p. 26.
22  World Development Movement, *Gunrunners' Gold*, London, World Development
    Movement, 1995, p. 15.

## Chapter 1

1   Keith Krause, *Arms and the State: Patterns of Military Production and Trade*, Cam-
    bridge, Cambridge University Press, 1992, p. 7.

2    SIPRI, *Yearbook of World Armaments and Disarmament*, London, Taylor and Francis, and Oxford, Oxford University Press; Arms Control and Disarmament Agency (ACDA), *World Military Expenditures and Arms Transfers*, Washington DC, US Government Printing Office; and International Institute for Strategic Studies, (IISS), *The Military Balance*, London, IISS.

3    Quoted in Paul Y. Hammond, David J. Louscher, Michael D. Salomone and N. Grahm, *The Reluctant Supplier: US Decision-Making for Arms Sales*, Cambridge, Mass., Oelgeschlager, Gunn and Hain, 1983, p. 1.

4    Christian Catrina, 'Main Directions of Research in the Arms Trade', in Robert E. Harkavy and Stephanie E. Neuman, *The Annals of the American Academy of Political and Social Science, The Arms Trade: Problems and Prospects in the Post-Cold War World*, Thousand Oaks, Ca., Sage, 1994, p. 204.

5    Lawrence Freedman, 'The Arms Trade: A Review', *International Affairs*, Vol. 55, No. 3, p. 433.

6    Michael T. Klare, 'The Political Economy of Arms Sales', *Bulletin of Atomic Scientists*, November 1976, p. 16.

7    Ian Anthony (Ed.), *Arms Export Regulations*, Oxford, Oxford University Press, 1991, p. 175.

8    Ministry of Defence, *Statement on the Defence Estimates*, Vol. II, various years.

9    National Audit Office, Report of the Comptroller and Auditor General, *Ministry of Defence: Support for Defence Exports*, House of Commons, Cmd 303, 1989, p. 1 and p. 6.

10    Frederic S. Pearson, 'Problems and Prospects of Arms Transfer Limitations Among Second-Tier Suppliers: The Cases of France, The United Kingdom and the Federal Republic of Germany', in Thomas Ohlson (Ed.), *Arms Transfer Limitations and Third World Security*, Oxford, Oxford University Press, 1989, p. 130.

11    National Audit Office, *op cit*, p. 9.

12    Christian Catrina, *Arms Transfers and Dependence*, New York, UN Institute for Disarmament Research, Taylor and Francis, 1988, p. 326.

13    See SIPRI, *The Arms Trade With the Third World*, London, Paul Elek, 1971, p. 36; *Staff Report to the Sub-Committee on Foreign Assistance of the Committee on Foreign Relations US Military Sales to Iran*, US Senate, US GPO, 1 July 1976, p. 53; and Anne Hessing Cahn *et al*, *Controlling Future Arms Trade*, New York, McGraw-Hill, 1977, p. 72.

14    Gloria Franklin, 'UK Arms Transfer Policy: A Defence', J. Simpson (Ed.), *The Control of Arms Transfers: Report of a FCO/BISA Seminar*, 23 September 1977, London, Arms Control and Disarmament Research Unit, FCO, 1977, p. 83.

15    Robert E. Harkavy, *The Arms Trade and International Systems*, Cambridge, Mass., Ballinger Publishing Company, 1975.

16    Robert E. Harkavy, 'The Changing International System and the Arms Trade' in *Annals*, *op cit*, p. 27.

17    Krause, *op cit*, p. 208.

18    Catrina, *Arms Transfers and Dependence*, *op cit*, p. 257.

19    Krause, *op cit*, p. 138.

20    Martin H. A. Edmonds, 'The Domestic and International Dimensions of British Arms Sales', in Cindy Cannizzo (Ed.), *The Gun Merchants: Politics and Policies of the Major Arms Suppliers*, Oxford, Pergamon Press, 1980, p. 76.

21    Ronald Matthews, 'Butter For Guns: The Growth of Under-the-Counter Trade', *The World Today*, May 1992, p. 90.

22    For an account see: World Development Movement, *Gunrunners' Gold*, London, World Development Movement, 1995, pp. 45–56.

23    National Audit Office, *op cit*, p. 29.

24    *Ibid*, p. 2.

25    See: *Hansard*, 16 June 1992, col. 460; and *Statement on the Defence Estimates*, Vol. II, Defence Statistics, 1991, cm no. 1559–II. The MOD asserted that arms sales offset domestic procurement by £300 million in 1995: see Minister for Defence Procurement, Roger Freeman, *Opening Statement* to the House of Commons Defence and Trade and

Industry Committees' Inquiry into Defence Procurement and Industrial Policy, 23 May 1995, London, MOD, Press Release. For discussions of the economic benefits of arms exports, see: J. Stanley and M. Pearton, *The International Trade in Arms*, London, Chatto and Windus, 1972; Frederick S. Pearson, 'Problems and Prospects', *op cit*; Anne Hessing Cahn, 'Arms Sales Economics: the Sellers' Perspectives', *Stanford Journal of International Studies*, Vol. 14, Spring 1979; Frederic S. Pearson, 'The Question of Control in British Defence Sales Policy', *International Affairs*, Vol. 59, No. 2, Spring 1983; Ron Smith, Anthony Humm and Jacques Fontanel, 'The Economics of Exporting Arms', *Journal of Peace Research*, Vol. 22, No. 3, 1985; World Development Movement, *Gunrunners' Gold*, 1995; Catrina, *Arms Transfers and Dependence*, *op cit*; and Trevor Taylor, 'Research Note: British Arms Exports and Research and Development Costs', *Survival*, Vol. 22, No. 6, December 1980.

26   Ron Smith, 'The Political Economy of Britain's External Relations', in Lawrence Freedman and Michael Clarke (Eds), *Britain in the World*, Cambridge University Press, 1991, p. 126; and Will Hutton, *The State We're In*, London, Vintage, 1996, p. 74.

27   Alan Clark, *Diaries*, London, Weidenfeld and Nicolson, 1993, p. 218.

28   Franklin, *op cit*, p. 84.

29   The Rt Hon. Sir Adam Butler, *Interview*, 24 April 1995.

30   Raimo Vayrynen, 'Economic and Social Consequences of Arms Transfers to the Third World', *Alternatives*, Vol. 6, No. 1, pp. 132–3.

31   SIPRI, *The Arms Trade with the Third World*, *op cit*, pp. 114–5.

32   Michael Brzoska and Thomas Ohlson, *Arms Transfers to the Third World 1971–1985*, Oxford, Oxford University Press for SIPRI, 1987, p. 74.

33   Andrew J. Pierre, *The Global Politics of Arms Sales*, Princeton, N.J., Princeton University Press, 1982, p. 101; and Krause, *op cit*, p. 144.

34   Export of Surplus War Material, Cmd 9676, 1955, quoted in Pierre, *Global Politics of Arms Sales, op cit*, p. 101.

35   Freedman, 'Arms Trade', *op cit*, p. 436.

36   Pierre, *Global Politics of Arms Sales, op cit*, pp. 101–2.

37   Michael Brecher, B. Steinberg and Janice Stein, 'A Framework for Research on Foreign Policy Behaviour', *Journal of Conflict Resolution*, Vol. 13, 1969, pp. 80–7.

38   Joseph Frankel, *National Interest*, London, Macmillan, 1970, p. 31.

39   Kenneth E. Boulding, 'National Images and International Systems', *Journal of Conflict Resolution*, Vol. 30, 1959, pp. 120–1.

40   See for example: R. Axelrod, *The Structure of Decision: The Cognitive Maps of Élites*, Princeton, Princeton University Press, 1976; Kenneth E. Boulding, *The Image*, Ann Arbor, Michigan University Press, 1976; R. Cottam, *Foreign Policy Motivation: A General Theory and a Case Study*, Pittsburgh, University of Pittsburgh Press, 1977; A. L. George, 'The "Operational Code": A Neglected Approach to the Study of Political Leaders and Decision-Making', *International Studies Quarterly*, Vol. 13, 1969, pp. 190–222; and R. Jervis, *Perception and Misperception in International Politics*, Princeton, Princeton University Press, 1976.

41   Hans J. Morgenthau, *Politics Among Nations*, Fourth Edition, New York, Knopf, 1967, p. 25.

42   Christopher Hill, 'The Historical Background: Past and Present in British Foreign Policy', in Michael Smith, Steve Smith and Brian White (Eds), *British Foreign Policy: Tradition, Change and Transformation*, London, Unwin Hyman, 1988, p. 28.

43   *Ibid*, p. 34.

44   *Ibid*, p. 31.

45   David Sanders, *Losing an Empire, Finding a Role: British Foreign Policy Since 1945*, London, Macmillan, 1990, p. 291.

46   Lord Franks quoted in P. Darby, *British Defence Policy East of Suez, 1947–68*, London, Oxford University Press, 1973, p. 22.

47   Margaret Thatcher, *The Downing Street Years*, London, HarperCollins, 1993, p. 8.

48   *Inquiry into Exports of Defence Equipment and Dual-Use Goods to Iraq*, Hearings in the presence of the Rt Hon. Lord Justice Scott, Evidence of Ian McDonald, Day 28, p. 59.

49   The Rt Hon. Alan Clark, *Interview*, 12 April 1995.
50   DESO official, *Interview*, August 1995.
51   Sir Anthony Parsons, Foreign Affairs Committee, *Current UK Policy Towards the Iran–Iraq Conflict*, Minutes of Evidence, 30 March 1988, Session 1987–88, House of Commons Papers, 279–II, p. 101, Q362.
52   The Rt Hon. Sir Adam Butler, *Interview*, 24 April 1995.
53   The Rt Hon. Alan Clark, *Interview*, 12 April 1995.
54   Michael Brzoska, *International Arms Transfers: the Nature and Dimensions of the Problem*, Hamburg, University of Hamburg, April 1990, p. 2.
55   *Inquiry, op cit*, Evidence of Lt Col. Glazebrook, Day 13, p. 89.
56   'Guidelines for the overseas promotion and supply of NBC defence equipment and the provision of NBC training', January 1988, quoted in *Inquiry*, Day 11, pp. 152–3.
57   *Guidelines for Sensitive Missile-Relevant Transfers*, 1992, p. 2, paragraph 2, provided by the Non-Proliferation and Defence Department, FCO.
58   *Inquiry, op cit,*, Evidence of Mr Vereker, Day 16, p. 96.
59   *Ibid*, p. 119; and Evidence of Michael Coolican, Day 59, p. 23.
60   Krause, *op cit*, p. 211.
61   Edmonds, 'Domestic and International Dimensions', *op cit*, p. 84.
62   Harkavy, 'Changing International System', *op cit*, p. 27.
63   Robin Luckham, 'Armaments, Underdevelopment and Demilitarisation in Africa', *Alternatives*, Vol. 6, No. 2, 1980, p. 180.
64   Andrew J. Pierre, 'Arms Sales the New Diplomacy', *Foreign Affairs*, Vol. 60, No. 2, Winter, 1981–82, p. 267.
65   *Inquiry, op cit*, Evidence of Eric Beston, Day 41, p. 12.
66   Douglas Hurd, FCO, Foreign Affairs Committee, *Overseas Arms Sales: Foreign Policy Aspects*, Minutes of Evidence, 4 March 1981, Session 1980–81, House of Commons Papers 41–vi, Q187.
67   Joanna Spear, 'Britain and Conventional Arms Transfer Restraint', in Mark Hoffman (Ed.), *UK Arms Control in the 1990s*, Manchester, Manchester University Press, 1990, p. 182.
68   *Inquiry, op cit*, Evidence of Eric Beston, Day 41, p. 9.
69   *Ibid*, Evidence of Michael Coolican, Day 59, p. 2.
70   Michael Blackwell, *Clinging to Grandeur: British Attitudes and Foreign Policy in the Aftermath of the Second World War*, London, Greenwood Press, 1993, p. 162.
71   Foreign Secretary, Foreign Affairs Committee, *Current UK Policy Towards the Iran–Iraq Conflict*, Minutes of Evidence, 27 January 1988, Session 1987–88, House of Commons Papers, 279–II, p. 7, Q2.
72   *Inquiry, op cit*, Evidence of David Gore-Booth, Day 21, p. 38.
73   *Ibid*, Evidence of John Major, MP, Day 55, p. 163.
74   Tim Sainsbury, MP, Letter to the Inquiry, 30 July 1993, quoted in *Inquiry*, Day 69, p. 25.
75   *Inquiry*, Evidence of Lord Trefgarne, Day 81, pp. 27 and 29.
76   *Ibid*, Evidence of Timothy Renton MP, Day 24, p. 146.
77   DESO official, *Interview*, August 1995.
78   The Rt Hon. Sir Adam Butler, *Interview*, 24 April 1995.
79   Interview with the Shah of Iran, *Business Week*, 24 January 1977, quoted in Chris Carr, 'Reverse Influence, Interdependence and the Relationship Between Supplier and Recipient in Arms Transfers' in Simpson, *op cit*, p. 23.
80   DTI, *Memorandum*, 5 December 1984, quoted in *Inquiry, op cit*, Day 7, p. 8.
81   *Foreign Policy Aspects of Overseas Arms Sales, op cit.*
82   Pearson, 'Question of Control', *op cit*, p. 215.
83   *Foreign Policy Aspects of Overseas Arms Sales, op cit.*
84   FCO, 'Guidelines for Defence Sales', 1984, quoted in *Inquiry, op cit*, Day 1, pp. 9–11.
85   *Inquiry*, Evidence of Lt Col. Glazebrook, Day 5, p. 21.
86   Head of Security, Overseas Commitments, *Memorandum*, 27 June 1990, quoted in *Inquiry*, Day 34, p. 40.
87   *Inquiry*, Evidence of Rob Young, Day 23, p. 84.

88  Alan Clark, Minister of State for Trade, *Hansard*, 19 April 1989.
89  MOD, Head of Security, Overseas Commitments, *Memorandum*, 27 June 1990, quoted in *Inquiry, op cit*, Day 34, p. 42.
90  Former Minister, *Interview*, 1995.
91  Saferworld Foundation, *Regulating Arms Exports: A Programme for the European Community*, Saferworld Foundation, 1991, p. 29.
92  Stanley and Pearton, *International Trade in Arms, op cit*, p. 16.
93  MOD, Head of DESS, *Memorandum*, 2 July 1990, quoted in *Inquiry, op cit*, Day 34, p. 54.
94  *Inquiry*, Evidence of Alan Clark, Day 49, pp. 14–15.
95  *Export of Goods (Control) Order 1987*, No. 2070, p. 5070.
96  *Inquiry, op cit*, Evidence of Anthony Steadman, Day 47, p. 162; and *Export of Goods (Control) Order, 1987*, No. 2070, Article 6, p. 5068.
97  *Inquiry*, Evidence of Eric Beston, Day 43, p. 38.
98  *Export of Goods (Control) Order 1987*, No. 2070, Article 5, p. 5068.
99  *Inquiry, op cit*, Evidence of Sir Adam Butler, Day 6, pp. 163–4.
100 *Ibid*, Evidence of Lt Col. Glazebrook, Day 12, pp. 82–3.
101 *Ibid*, p. 83.
102 *Ibid*, Evidence of Alan Clark, Day 50, p. 31.
103 Pearson, 'Problems and Prospects', *op cit*, p. 134.
104 *Inquiry, op cit*, Evidence of Anthony Steadman, Day 46, p. 94.
105 Edmonds, 'Domestic and International Dimensions', *op cit*, p. 95.
106 *Inquiry, op cit*, Evidence of Michael Coolican, Day 59, pp. 150–1.
107 Export Control Organisation, DTI, *A Guide to Export Controls*, London, February 1995, para. 3.3, quoted in Paul Cornish, *The Arms Trade and Europe*, Chatham House Papers, The Royal Institute of International Affairs, 1995, p. 30.
108 Timothy Renton, Minister of State, FCO, *Hansard*, 19 February 1987.
109 *Inquiry, op cit*, Sir Richard Scott, Day 37, p. 111.
110 *Ibid*, Evidence of Michael Coolican, Day 59, p. 77.
111 Michael Coolican, Head of Export Control Organisation, DTI, Trade and Industry Committee, *Exports to Iraq*, Minutes of Evidence, 26 November 1991, Session 1990–91, House of Commons Papers 86–i, p. 4, Q30.


Chapter 2
1  The Rt Hon. Sir Richard Scott, the Vice-Chancellor, *Report of the Inquiry into the Export of Defence Equipment and Dual-Use Goods to Iraq and Related Prosecutions*, House of Commons Papers 115, 1996, K8.15–16, pp. 1805–6.
2  See (the founding texts): Graham T. Allison, *Essence of Decision: Explaining the Cuban Missile Crisis*, Boston, Mass., Little Brown and Co., 1971; and Morton H. Halperin, *Bureaucratic Politics and Foreign Policy*, Washington DC, Brookings Institution, 1974.
3  Some of the best critiques of the bureaucratic politics and organizational process models are: J. P. Cornford, 'Review Article: The Illusion of Decision', *British Journal of Political Science*, Vol. 4, 1974, pp. 231–43; Lawrence Freedman, 'Logic, Politics and Foreign Policy Processes: A Critique of the Bureaucratic Politics Model', *International Affairs*, Vol. 52, 1976, pp. 434–49; and Stephen D. Krasner, 'Are Bureaucracies Important? (Or Allison in Wonderland)', *Foreign Policy*, Vol. 7, 1972, pp. 159–79.
4  William Wallace, *The Foreign Policy Process in Britain*, London, Royal Institute of International Affairs, 1975, p. 9.
5  Steve Smith, 'Foreign Policy Analysis', in Lawrence Freedman and Michael Clarke (Eds), *Britain in the World*, Cambridge, Cambridge University Press, 1991, p. 68.
6  Alan Clark, *Diaries*, London, Weidenfeld and Nicolson, 1993, p. 64.
7  Martin H. A. Edmonds, 'The Domestic and International Dimensions of British Arms Sales', in Cindy Cannizzo (Ed.), *The Gun Merchants: Politics and Policies of the Major Arms Suppliers*, Oxford, Pergamon Press, 1980, p. 80.
8  *Inquiry into Exports of Defence Equipment and Dual-Use Goods to Iraq*, Hearings in

the presence of the Rt Hon. Lord Justice Scott, Evidence of Sir Robin Butler, Day 62, p. 5.

9   J. J. Richardson and A. G. Jordan, *Governing Under Pressure, The Policy Process in a Post-Parliamentary Democracy*, Oxford, Martin Robertson, 1979, and *British Politics and the Policy Process, An Arena Approach*, London, Allen and Unwin, 1987. For surveys of models of British politics, see Andrew Gamble, 'Theories of British Politics', *Political Studies*, Vol. 38, 1990, pp. 404–20, or P. Dunleavy and B. O'Leary, *Theories of the State*, London, Macmillan, 1987.

10  *Inquiry, op cit*, Evidence of Sir Robin Butler, Day 62, p. 82.

11  *Ibid*, Evidence of John Major, MP, Day 55, pp. 174–5.

12  *Ibid*, Evidence of Lord Howe, Day 54, p. 27.

13  *Ibid*, Evidence of Lord Trefgarne, Day 82, p. 206.

14  *Ibid*, Evidence of John Major, MP, Day 55, p. 88.

15  *Ibid*, Evidence of Ian McDonald, Day 28, p. 107.

16  *Ibid*, Evidence of Timothy Renton, MP, Day 24, p. 9.

17  *Ibid*, Evidence of Ian McDonald, Day 34, p. 5.

18  *Ibid*.

19  H. Heclo and A. Wildavsky, *The Private Government of Public Money*, London, Macmillan Press, 1974, p. 283.

20  *Ibid*, p. 80.

21  *Inquiry, op cit*, Evidence of Alan Barrett, Day 36, p. 106.

22  *Ibid*, Evidence of Stephen Lamport, Day 32, p. 88.

23  *Ibid*, Evidence of Eric Beston, Day 44, p. 62.

24  *Ibid*, Evidence of Douglas Hurd, MP, Day 57, pp. 24–5.

25  *Ibid*, Evidence of Mr Vereker, Day 16, p. 101.

26  Roger Hilsman, *To Move a Nation*, New York, Doubleday, 1967, p. 197.

27  *Inquiry, op cit*, Evidence of Baroness Thatcher, Day 48, p. 12.

28  *Ibid*, Evidence of Lord Trefgarne, Day 81, p. 106.

29  *Ibid*, Evidence of Sir Stephen Egerton, Day 11, p. 51.

30  *Ibid*, Evidence of Eric Beston, Day 42, p. 95.

31  'Questions of Procedure for Ministers', *Memorandum* from the Cabinet Secretary, Sir Robert Armstrong, quoted in *Inquiry*, Day 48, pp. 19–20.

32  *Inquiry*, Evidence of Alan Clark, Day 50, p. 145.

33  *Ibid*, Evidence of David Gore-Booth, Day 21, p. 46.

34  *Ibid*, Evidence of Timothy Renton, MP, Day 24, pp. 14–15.

35  'Osmotherly Rules', quoted in *Inquiry*, Day 62, p. 75 and pp. 113–14.

36  *Inquiry*, Evidence of Sir Robin Butler, Day 62, pp. 101–2.

37  Trade and Industry Committee, *Second Report, Exports to Iraq: Project Babylon and Long Range Guns*, Session 1991–92, House of Commons Papers 86.

38  *Inquiry, op cit*, Evidence of Nicholas Bevan, Day 56, p. 71.

39  Edmonds, *op cit*, pp. 80–1.

40  Denzil Davies, Labour Party Spokesman on Defence, 29 March and 17 April 1985, quoted in Campaign Against Arms Trade, *Iran and Iraq: Arms Sales and the Gulf War*, London, Campaign Against Arms Trade, 1986; and Derek Fatchett, Labour Party Spokesman on Foreign Affairs, Speech to Lobster Readers' Meeting, Leeds, 24 February 1996.

41  *Inquiry, op cit*, Evidence of Alan Clark, Day 49, p. 97.

42  *Ibid*, Sir Richard Scott, Day 20, p. 139.

43  Frederick S. Pearson, 'Problems and Prospects of Arms Transfer Limitations Among Second-Tier Suppliers: The Cases of France, the United Kingdom and the Federal Republic of Germany', in Thomas Ohlson (Ed.), *Arms Transfer Limitations and Third World Security*, Oxford, Oxford University Press, 1988, p. 129.

44  National Audit Office, Report of the Comptroller and Auditor General, *Ministry of Defence: Support for Defence Exports*, House of Commons, Cmd 303, 1989, p. 5.

45  Sir Ronald Ellis, Head of Defence Sales, Lecture given at the Royal United Services Institute for Defence Studies, 18 October 1978, *RUSI Journal*, Vol. 124, 1979, p. 6.

46  Frederic S. Pearson, 'The Question of Control in British Defence Sales Policy', *Interna-*

*tional Affairs*, Vol. 59, No. 2, Spring 1983, p. 235.
47  *Inquiry, op cit*, Evidence of Sir Stephen Egerton, Day 11, p. 76.
48  *Ibid*, Evidence of Eric Beston, Day 41, p. 18.
49  *Ibid*, Evidence of Lord Howe, Day 54, p. 35.
50  *Ibid*, Evidence of Rob Young, Day 23, p. 214.
51  *Ibid*.
52  *Ibid*, Evidence of Lt Col. Glazebrook, Day 5, pp. 43–4.
53  *Ibid*, Evidence of Douglas Hurd, MP, Day 57, p. 15.
54  *Ibid*, Evidence of Christopher Sandars, Day 3, pp. 7–8.
55  *Ibid*, Evidence of Ian McDonald, Day 28, p. 163.
56  *Ibid*, Evidence of Christopher Sandars, Day 3, p. 22 and pp. 4–5.
57  *Ibid*, p. 11.
58  *Ibid*, Evidence of Ian McDonald, Day 28, p. 161.
59  *Ibid*, Evidence of Michael Coolican, Day 59, p. 9.
60  *Ibid*, Evidence of Eric Beston, Day 41, p. 164.
61  *Ibid*, Evidence of Alan Barrett, Day 39, p. 11.
62  *Ibid*, Evidence of William Patey, Day 14, pp. 4–5, and Rob Young, Day 23, p. 134.
63  *Ibid*, Evidence of Eric Beston, Day 41, p. 181.
64  *Ibid*, p. 195.
65  *Report, op cit*, K3.3, p. 1769.
66  *Inquiry, op cit*, Evidence of Eric Beston, Day 41, p. 152.
67  *Ibid*, p. 18.
68  *Ibid*, Evidence of Paul Channon, Day 7, p. 148.
69  *Ibid*, Evidence of Michael Coolican, Day 59, p. 20.
70  *Ibid*, Evidence of Eric Beston, Day 41, pp. 17–18 and Day 42, p. 116.
71  *Ibid*, Day 41, p. 159.
72  *Report, op cit*, K3.2, p. 1768.
73  *Inquiry, op cit*, Evidence of Michael Coolican, Day 59, p. 3.
74  *Ibid*, Evidence of Michael Petter, Day 58, p. 107.
75  *Ibid*, Evidence of Michael Coolican, Day 59, p. 19.
76  *Ibid*, Evidence of Eric Beston, Day 41, p. 117.
77  *Ibid*, Evidence of Alan Barrett, Day 38, pp. 73–4.
78  *Ibid*, Evidence of Lord Trefgarne, Day 82, p. 80.
79  *Ibid*, Evidence of Anthony Steadman, Day 47, pp. 91–2.
80  *Ibid*, Evidence of Eric Beston, Day 41, pp. 91–2.
81  See: *Report, op cit*, D6.55–D6.72, pp. 587–94.
82  *Inquiry, op cit*, Evidence of Eric Beston, Day 41, pp. 83–4.
83  *Ibid*, Sir Richard Scott, Day 47, p. 99.
84  *Ibid*, Evidence of Michael Coolican, Day 59, p. 10.
85  *Ibid*, p. 11.
86  *Ibid*, Evidence of Alan Barrett, Day 36, p. 22.
87  *Ibid*, Evidence of Eric Beston, Day 41, p. 9.
88  *Ibid*, Evidence of Ian McDonald, Day 28, p. 14.
89  *Ibid*, Evidence of Simon Fuller, Day 31, p. 164.
90  *Ibid*, Evidence of Eric Beston, Day 43, p. 38.
91  *Ibid*, Day 41, p. 196.
92  *Ibid*, Day 43, p. 113.
93  *The Guardian*, 14 June 1995.
94  *Memorandum* by the Department of Trade and Industry, December 1990, Trade and Industry Committee, *Exports to Iraq*, Memoranda of Evidence, 17 July 1991, Session 1990–91, House of Commons Papers 607, p. 6.
95  *Inquiry, op cit*, Day 39, p. 74.
96  Pearson, 'Question of Control', *op cit*, p. 215.
97  *Inquiry, op cit*, Evidence of Alan Barrett, Day 36, p. 2.
98  *Ibid*, Evidence of Christopher Sandars, Day 3, p. 160.
99  *Ibid*, Evidence of Lt Col. Glazebrook, Day 13, p. 99.
100 *Ibid*, Evidence of Eric Beston, Day 41, p. 43.

101  J. Stanley and M. Pearton, *The International Trade in Arms*, London, Chatto and Windus, 1972, p. 93.

102  Andrew .J Pierre, *The Global Politics of Arms Sales*, Princeton, New Jersey, Princeton University Press, 1982, p. 103.

103  J. McGowan, 'The Future of UK Defence Exports', *Seaford House Papers*, London, Royal College of Defence Studies, p. 140.

104  *Inquiry, op cit*, Evidence of Christopher Sandars, Day 3, p. 149.

105  *Ibid*, Evidence of Ian McDonald, Day 28, pp. 124–5.

106  *Ibid*, Evidence of Christopher Sandars, Day 3, p. 149.

107  National Audit Office, *op cit*, p. 13 and p. 2.

108  *Ibid*.

109  *Inquiry, op cit*, Evidence of Alan Barrett, Day 36, pp. 17–18.

110  *Ibid*, Sir Richard Scott, Day 7, p. 144.

111  R. Anderson, Defence Sales Organisation, Expenditure Committee, Minutes of Evidence, 4 November 1975, Session 1975–75, House of Commons Papers 155–I, p. 214, Q800.

112  The complete narrative is given in *Inquiry, op cit*, Day 34, pp. 154–216.

113  *Inquiry*, Evidence of Ian McDonald, Day 34, p. 161 and p. 157.

114  Embassy, Baghdad, *Telex* to FCO, 1 May 1989, quoted in *Inquiry*, Day 34, p. 197.

115  FCO, *Letter* to the DESO, 22 March 1989, quoted in *Inquiry*, Day 34, p. 171.

116  FCO, Minister of State, *Letter* to Minister for Defence Procurement, 28 April 1989, quoted in *Inquiry*, Day 34, pp. 182–3.

117  *Inquiry*, Evidence of Ian McDonald, Day 34, p. 175.

118  Embassy, Baghdad, *Telex* to FCO, 11 July 1989, quoted in *Inquiry*, Day 34, pp. 199–200.

119  *Ibid*, p. 203.

120  *Inquiry*, Evidence of Ian McDonald, Day 34, p. 203.

121  *Ibid*, pp. 155–6, p. 157.

122  *Ibid*, Evidence of Sir Stephen Egerton, Day 11, p. 75.

123  Ellis, *op cit*, p. 5.

124  *Inquiry, op cit*, Evidence of Lord Trefgarne, Day 81, p. 73.

125  *Ibid*, Evidence of Lt Col. Glazebrook, Day 4, p. 93.

126  *Ibid*, Evidence of Ian Blackley, Day 20, p. 179.

127  *Ibid*, Evidence of Lt Col. Glazebrook, Day 4, p. 94.

128  *Ibid*, Day 12, p. 24 and p. 23.

129  *Ibid*, Day 5, p. 76.

130  *Ibid*, Day 4, pp. 58–9.

131  *Ibid*, p. 58.

132  *Ibid*, pp. 66–7.

133  *Ibid*, pp. 64–5.

134  *Ibid*, Evidence of Ian McDonald, Day 28, p. 11.

135  *Ibid*, Evidence of Christopher Sandars, Day 3, p. 15.

136  *Ibid*, Evidence of Ian McDonald, Day 28, p. 11.

137  *Ibid*, Evidence of Christopher Sandars, Day 3, pp. 163–4.

138  *Ibid*, p. 160.

139  *Ibid*, Evidence of Alan Barrett, Day 36, p. 8.

140  *Ibid*, Evidence of Lt Col. Glazebrook, Day 12, p. 118.

141  *Ibid*, Evidence of Ian McDonald, Day 29, p. 44.

142  *Ibid*, Evidence of Nicholas Bevan, Day 56, p. 42.

143  *Ibid*, Evidence of Ian McDonald, Day 28, p. 151.

144  MOD, MinAF, *Memorandum* to MinDP, 22 December 1986, quoted in *Inquiry*, Day 5, pp. 18–19.

145  *Inquiry*, Evidence of Lt Col. Glazebrook, Day 5, p. 20.

146  *Ibid*, Evidence of Alan Barrett, Day 36, p. 64.

147  *Ibid*, Evidence of Ian McDonald, Day 34, p. 22 and p. 21.

148  *Ibid*, Evidence of Alan Barrett, Day 36, pp. 67–8.

149  Lt Col. Glazebrook's notes of the MOD Working Group Meeting, 7 March 1990, quoted in *Inquiry*, Day 5, pp. 24–5.

150 *Inquiry*, Evidence of Alan Clark, Day 49, p. 2.
151 *Ibid*, Evidence of Lord Trefgarne, Day 81, pp. 12–13.
152 *Ibid*, Evidence of Alan Clark, Day 49, p. 165.
153 The Rt Hon. Sir Adam Butler, *Interview*, 24 April 1995.
154 *Inquiry, op cit*, Evidence of Sir Adam Butler, Day 6, pp. 1–2.
155 *Ibid*, Evidence of Christopher Sandars, Day 3, p. 156.
156 *Ibid*, Evidence of Alan Barrett, Day 36, pp. 6–7.
157 *Ibid*, Evidence of Sir Adam Butler, Day 6, p. 3.
158 The Rt Hon. Sir Adam Butler, *Interview*, 24 April 1995.
159 Pearson, 'Question of Control', *op cit*, p. 237.
160 The Rt Hon. Sir Adam Butler, *Interview*, 24 April 1995.
161 MOD, AUS Commitments, *Minute*, 27 June 1990, quoted in *Inquiry, op cit*, Day 34, pp. 40–41.
162 MOD, Head of Security (Overseas Commitments), *Minute*, 29 June 1990, quoted in *Inquiry*, Day 34, pp. 47–8.
163 *Ibid*, quoted in *Inquiry*, Day 34, p. 45.
164 Head of Export Control Organisation, Trade and Industry Committee, *op cit*, Minutes of Evidence, 26 November 1991, House of Commons Papers 86-i, p. 9, Q90.
165 Pearson, 'Question of Control', *op cit*, p. 227.
166 *Inquiry, op cit*, Evidence of Mark Higson, Day 17, p. 21; and Evidence of Charles Haswell, Day 33, p. 17.
167 The Rt Hon. Alan Clark, *Interview*, 12 April 1995.
168 *Inquiry, op cit*, Evidence of Mr Vereker, Day 16, p. 15 and pp. 12–14.
169 Clark, *Diaries, op cit*, p. 161.
170 *Inquiry, op cit*, Evidence of Stephen Day, Day 1, p. 194.
171 Clark, *Diaries, op cit*, p. 56.
172 *Inquiry, op cit*, Evidence of David Mellor, MP, Day 25, p. 58.
173 *Ibid*, Evidence of Douglas Hurd MP, Day 57, p. 15.
174 National Audit Office, *op cit*, p. 4.
175 *Ibid*, pp. 29–30.
176 *Inquiry, op cit*, Evidence of Mr Bryars, Day 2, p. 37.
177 *Ibid*, Evidence of Sir Stephen Egerton, Day 11, p. 96.
178 *Ibid*, Evidence of Mr Fellgett, Day 15, p. 13.
179 *Ibid*, Evidence of Lt Col. Glazebrook, Day 12, p. 42.
180 *Ibid*, Evidence of Alan Clark, Day 50, p. 31.
181 DTI, Anthony Steadman, *Memorandum* to Customs and Excise, 28 December 1990, (000416)*.
    (* Numbered documents refer to those released to Paul Henderson and passed to the author.)
182 *Inquiry, op cit*, Evidence of Mr Vereker, Day 16, pp. 6–7, p. 144.
183 FCO, *Memorandum*, 27 June 1990, quoted in *Inquiry*, Day 22, p. 165.
184 *Ibid*, Evidence of Lt Col. Glazebrook, Day 4, p. 140; and Evidence of Christopher Sandars, Day 3, p. 20.
185 *Ibid*, Evidence of Alan Barrett, Day 38, p. 18.
186 MOD, *Memorandum*, 20 July 1988, quoted in *Inquiry*, Day 20, p. 60.
187 *Inquiry*, Evidence of Michael Coolican, Day 59, p. 37.
188 *Ibid*, Evidence of Lord Trefgarne, Day 81, p. 101.
189 *Ibid*, Evidence of Lady Thatcher, Day 48, p. 59.
190 Head of Treasury Division, Defence Management 1, *Memorandum*, 26 June 1990, quoted in *Inquiry*, Day 15, p. 25.
191 Michael Brzoska and Thomas Ohlson, *Arms Transfers to the Third World 1971–85*, Oxford, Oxford University Press for SIPRI, 1987, p. 76.
192 *Inquiry, op cit*, Day 12, p. 128.
193 *Ibid*, Evidence of Eric Beston, Day 41, p. 199.
194 Theodore Sorenson, *Decision-Making in the White House: the Olive Branch or the Arrows*, New York, Columbia University Press, 1963, p. 44.
195 Keith Krause, *Arms and the State: Patterns of Military Production and Trade*, Cam-

bridge, Cambridge University Press, 1992, p. 143.

196  *Inquiry, op cit*, Evidence of Stephen Lamport, Day 32, p. 57.

197  *Ibid*, Evidence of Michael Petter, Day 58, p. 2.

198  *Ibid*, Evidence of Alan Collins, Day 10, p. 57.

199  *Ibid*, Evidence of David Gore-Booth, Day 21, p. 12.

200  *Ibid*, Evidence of Paul Channon, Day 7, pp. 9–10 and p. 18.

201  Brzoska and Ohlson, *op cit*, p. 75; and see also, Pearson, 'Question of Control', *op cit*, p. 227.

202  FCO, *Memorandum*, 20 July 1988, quoted in *Inquiry, op cit*, Day 20, p. 60.

203  Joanna Spear, 'Britain and Conventional Arms Transfer Restraint', in Mark Hoffman (Ed.), *UK Arms Control in the 1990s*, Manchester, Manchester University Press, 1990, p. 176.

204  Pearson, 'Question of Control', *op cit*, p. 227.

205  *Ibid*, p. 237.

206  The Rt Hon. Sir Adam Butler, *Interview*, 24 April 1995.

207  Krause, *op cit*, p. 143.

208  Ivan Lawrence, MP, *Current UK Policy Towards the Iran/Iraq Conflict*, Foreign Affairs Committee, Minutes of Evidence, 30 March 1988, Session 1987–88, House of Commons Papers 279–vi, p. 107, Q391.

209  *Inquiry, op cit*, Evidence of Mr Fellgett, Day 15, pp. 28–9.

210  National Audit Office, *op cit*, p. 26.

211  Pierre, *op cit*, p. 103.

212  R. Steel in Morton Halperin, *Readings in American Foreign Policy*, Boston, Mass., Little, Brown and Co, 1973, p. 203.

## Chapter 3

1   The Rt Hon. Sir Richard Scott, the Vice-Chancellor, *Report of the Inquiry into the Export of Defence Equipment and Dual-Use Goods to Iraq and Related Prosecutions*, House of Commons Papers, 115, 1996, D2.427, p. 367.

2   FCO, Middle East Department, *Memorandum* to the Defence Department, 13 October 1987, (000005)*.
    (* Numbered documents refer to those released to Paul Henderson and passed to the author.)

3   Three executives of Ordnance Technologies and a director of a transport company pleaded guilty at Reading Crown Court to charges of evasion of the prohibition on exportation contrary to the Customs and Excise Management Act in February 1992. A successful appeal was mounted on the grounds that the Government knew that the fuse assembly line for Jordan was in fact destined for Iraq.

4   *Inquiry into Exports of Defence Equipment and Dual-Use Goods to Iraq*, Meeting between the Rt Hon. Lord Justice Scott and the Prosecuting Counsel in the Ordnance Technologies case, 16 June 1993, p. 11.

5   *Ibid*.

6   *Hansard*, 29 October 1985, Vol. 84, col. 454.

7   *Inquiry into Exports of Defence Equipment and Dual-Use Goods to Iraq*, Hearings in the presence of the Rt Hon. Lord Justice Scott, Evidence of Lord Howe, Day 54, p. 17.

8   See George P. Politakis, 'Variations on a Myth: Neutrality and the Arms Trade', *German Yearbook of International Law*, Vol. 35, 1992, pp. 435–506.

9   Douglas Hurd, Minister of State, FCO, *Overseas Arms Sales: Foreign Policy Aspects*, Minutes of Evidence, 4 March 1981, Session 1980–81, House of Commons Papers 41–iv, Q194.

10  Mr Alan Clark, Minister of State for Defence Procurement, *Hansard*, 2 May 1990, col. 603.

11  *Inquiry, op cit*, Evidence of Sir Adam Butler, Day 6, p. 6.

12  *Ibid*, Evidence of Sir Stephen Egerton, Day 11, pp. 82–3.

13  *Report of the Congressional Committees Investigating the Iran-Contra Affair*, H. Rept. No. 100–433, S.Rept No. 100–216, 100th Congress, First Session, Washington D.C., US Government Printing Office, 1987.

14   *The Guardian*, 23 February 1984.
15   *The Sunday Times*, 19 August 1984.
16   United States, Department of State, 'Iran–Iraq War (Discussion Points for Visit of Prime Minister Thatcher)', *National Security Archive*, Document 64218, 12 December 1984.
17   *Inquiry, op cit*, Evidence of Christopher Sandars, Day 3, p. 52 and p. 54.
18   *Ibid*, Evidence of Alan Collins, Day 10, p. 85.
19   FCO, Minister of State, *Letter* to MinDP, 2 November 1984, quoted in *Inquiry,* Day 6, p. 18.
20   MOD, MinDP, *Minute*, November 1984, quoted in *Inquiry,* Day 6, p. 40.
21   FCO, *Memorandum*, 24 October 1984, quoted in *Inquiry*, Day 11, pp. 27–28.
22   *Inquiry*, Evidence of Sir Richard Luce, Day 1, p. 4.
23   *Ibid*, pp. 7–8.
24   *Report, op cit*, D1.5–6, p. 153.
25   *Inquiry, op cit*, Evidence of Sir Richard Luce, Day 1, pp. 32–3.
26   *Ibid*, Evidence of Stephen Day, Day 1, p. 160.
27   FCO, Minister of State, *Memorandum*, 'Ban on sales of all defence equipment to Iran and Iraq', 26 July 1994, quoted in *Inquiry*, Day 1, pp. 24–5.
28   Foreign Secretary, *Minute*, 4 December 1984, quoted in *Inquiry*, Day 1, pp. 76–7.
29   *Inquiry,* Evidence of Stephen Day, Day 1, p. 177.
30   Trade and Industry Committee, *Exports to Iraq*, Memorandum from the DTI, Annex E, Memoranda of Evidence, Session 1990–91, House of Commons Papers 607, pp. 43–48.
31   *Report, op cit*, D3.187, p. 466.
32   *Ibid*, D2.3, p. 215.
33   *Ibid*, D1.15, p. 156.
34   *Ibid*, D2.56, p. 234 and D2.60, p. 235.
35   *Ibid*, D3.67, p. 400.
36   *Annex to Brief for Minister for Defence Procurement for the Foreign Secretary's Meeting*, 19 July 1990, (000410).
37   *Inquiry, op cit*, Evidence of Christopher Sandars, Day 3, pp. 43–4.
38   *Ibid*, Evidence of Lt Col. Glazebrook, Day 4, p. 54.
39   International Institute for Strategic Studies, *The Military Balance*, London, IISS, 1980–81, pp. 42–3.
40   *Inquiry, op cit*, Evidence of Sir Richard Luce, Day 1, p. 71.
41   *Ibid*, Evidence of Lt Col. Glazebrook, Day 4, pp. 74–5.
42   FCO, Minister of State, *Letter* to Minister for Trade, 27 April 1989, (000169–70).
43   *Inquiry, op cit*, Sir Richard Scott, Day 6, pp. 153–4.
44   *Ibid*, Day 28, p. 42.
45   *Ibid*, Evidence of Sir Stephen Egerton, Day 11, p. 71.
46   *Ibid*, Evidence of Ian McDonald, Day 28, p. 159.
47   FCO, Middle East Department, Mr Haskell, Comments upon a *Memorandum*, 24 October 1984, quoted in *Inquiry*, Day 11, pp. 27–8.
48   FCO, MED, Under-Secretary, *Memorandum*, 24 October 1984, quoted in *Inquiry*, Day 11, pp. 27–8.
49   MOD, *Record of the MOD WG meeting*, 13 December 1984, quoted in *Inquiry*, Day 4, p. 29.
50   *Minute* of the Ministerial meeting to review the Guidelines, 19 June 1986, quoted in *Inquiry*, Day 4, p. 36.
51   *Report, op cit*, D3.2, p. 372.
52   *Ibid*, D3.102, p. 415.
53   Foreign Secretary, 'The Economic Consequences of an End to the Iran–Iraq Conflict', 31 August 1988, quoted in *Inquiry, op cit*, Day 48, p. 40.
54   Private Secretary to the Prime Minister, *Letter* to Private Secretary to the Foreign Secretary, 2 September 1988, quoted in *Report, op cit*, D3.14, p. 377.
55   FCO, MED, *Paper*, 9 September 1988, quoted in *Report*, D3.16–17, pp. 377–8.
56   Private Secretary to the Foreign Secretary, *Note* to FCO officials and Minister of State, 22 September 1988, quoted in *Report*, D3.19, p. 379.

57 Private Secretary to the Foreign Secretary, *Note*, 28 October 1988, quoted in *Report*, D3.21, p. 380.
58 FCO, Minister of State, *Letter* to Minister for Trade, 14 November 1988, quoted in *Report*, D3.24, pp. 381–2.
59 FCO, Private Secretary to Minister of State, *Letter* to Private Secretaries to Minister for Trade and MinDP, 23 December 1988, quoted in *Report*, D3.36, pp. 387–8.
60 FCO, *Summary Record of the IDC*, 14 March 1989, (000148).
61 MOD, DESS 2, *Note* to MODWG, 6 February 1989, quoted in *Report, op cit*, D3.60, p. 397.
62 FCO, Minister of State, *Letter* to Minister for Trade, 27 April 1989, quoted in *Report*, D3.83, pp. 406–7.
63 FCO, *Submission*, 'UK Policy on Iran/Iraq', to the Foreign Secretary, 28 April 1989, quoted in *Report*, D3.85, p. 407.
64 Foreign Secretary, *Minute* to the Prime Minister, 3 March 1989, quoted in *Report*, D3.67, pp. 399–400.
65 *Report*, D3.102, p. 415.
66 See, for example, *Inquiry, op cit*, Evidence of Alan Clark, Day 49, pp. 144–7.
67 *Inquiry*, Evidence of Lord Howe, Day 54, pp. 133–4.
68 *The Guardian*, 15 April 1996.
69 See: *Report, op cit*, D3.165, pp. 458–9.
70 *Inquiry, op cit*, Day 30, p. 44.
71 *Ibid*, Evidence of Timothy Renton, MP, Day 24, p. 56.
72 Private Secretary to the Prime Minister, *Note of Meeting between the Prime Minister and Tariq Aziz*, 4 December 1985, quoted in *Inquiry*, Day 24, p. 55.
73 FCO, *Summary Record of the IDC*, 10 December 1985, quoted in *Inquiry*, Day 10, p. 108.
74 *Inquiry*, Evidence of Lt Col. Glazebrook, Day 4, pp. 20–1.
75 DTI, Director Export Licensing Branch, *Letter* to DESS 2, 13 January 1988, (000024).
76 *Inquiry, op cit*, Evidence of Lt Col. Glazebrook, Day 4, p. 96.
77 *Ibid*, pp. 68–9.
78 *Ibid*, Evidence of Stephen Day, Day 1, p. 162.
79 *Ibid*, Evidence of Lt Col. Glazebrook, Day 4, p. 8.
80 *Ibid*, Day 5, pp. 32–3.
81 *Ibid*, pp. 43–4.
82 *Ibid*, Evidence of Lord Howe, Day 54, p. 20.
83 *Ibid*, Evidence of Alan Clark, Day 49, p. 3 and p. 7.
84 *Ibid*, Evidence of Lord Howe, Day 54, p. 25.
85 *Ibid*, Evidence of Michael Petter, Day 58, p. 27.
86 FCO, *Submission* to the Minister of State, 26 January 1989, quoted in *Inquiry*, Day 26, pp. 156–7.
87 FCO, *Briefing* for the Foreign Secretary, 23 June 1989, quoted in *Inquiry*, Day 21, pp. 141–2.
88 Treasury, *Memorandum*, 25 July 1989, quoted in *Inquiry*, Day 15, p. 10.
89 *Hansard*, 12 April 1984, cols. 339–340.
90 *Inquiry, op cit*, Evidence of Sir Adam Butler, Day 6, p. 27.
91 FCO, *Minute*, 26 November 1986, quoted in *Inquiry*, Day 18, pp. 94–5.
92 FCO, *Memorandum*, 7 August 1987, quoted in *Inquiry*, Day 18, pp. 102–3.
93 FCO, *Summary Record of the IDC*, 14 March 1989, (000148).
94 *Ibid*, 23 March 1989, (000161).
95 FCO, internal guidelines for considering arms export applications, 2 April 1984, quoted in *Inquiry, op cit*, Day 1, p. 11.
96 *Inquiry*, Evidence of Mark Higson, Day 17, p. 52.
97 *Ibid*, Evidence of Lt Col. Glazebrook, Day 4, p. 30.
98 *Report, op cit*, D1.163, p. 210.
99 *Minute* of a Ministerial Meeting, 19 June 1986, quoted in *Inquiry*, Day 4, p. 37.
100 MODWG, Lt Col. Glazebrook, *Memorandum* to DESS 2, 14 February 1989, quoted in *Inquiry*, Day 4, pp. 44–5.

101 MODWG, Lt Col. Glazebrook, *Minute*, 14 February 1898, quoted in *Inquiry*, Day 4, pp. 44–5.
102 Mr John Goulden CMG, Assistant Under-Secretary of State at the Foreign and Commonwealth Office, Trade and Industry Committee, *Exports to Iraq*, Minutes of Evidence, 28 January 1992, Session 1991–92, House of Commons Papers, 86–viii, p. 281; and *Defence Sales to Iran and Iraq: Proposed Revision*, 20 December 1988, (000111).
103 FCO, Minister of State, *Letter* to Minister for Trade, 27 April 1989, (000169–70).
104 *Note* of the Foreign Secretary's meeting with the Prime Minister, 26 July 1990, quoted in *Inquiry, op cit*, Day 27, p. 185.
105 *Note* of the Ministerial Meeting, 19 July 1990, quoted in *Inquiry*, Day 30, p. 194.
106 FCO, *Telex* to Embassy, Washington, 19 July 1990, quoted in *Inquiry*, Day 30, p. 194.
107 *Inquiry*, Evidence of Stephen Day, Day 1, pp. 151–2.
108 *Ibid*, Evidence of Lord Howe, Day 54, p. 9.
109 *Report, op cit*, D1.165, p. 211.

**Chapter 4**

1  *Inquiry into Exports of Defence Equipment and Dual-Use Goods to Iraq*, Hearings in the presence of Lord Justice Scott, Evidence of Sir Richard Luce, Day 1, p. 7.
2  Foreign Affairs Committee, *Second Report: Current UK Policy Towards The Iran–Iraq Conflict*, Session 1987–88, House of Commons Papers, 279–I, p. xix.
3  FCO, *Submission*, 5 November 1984, quoted in the Rt Hon. Sir Richard Scott, the Vice-Chancellor, *Report of the Inquiry into the Export of Defence Equipment and Dual-Use Goods to Iraq and Related Prosecutions*, House of Commons Papers 115, 1996, D1.40, p. 166.
4  *Inquiry, op cit*, Evidence of Stephen Day, Day 1, p. 164.
5  *Ibid*, Evidence of Ian Blackley, Day 20, p. 6.
6  The Rt Hon. Sir Adam Butler, *Interview*, 24 April 1995.
7  Meeting between Minister of State for Foreign and Commonwealth Affairs and MinDP, 24 October 1984, quoted in *Inquiry, op cit*, Day 1, p. 35.
8  *Inquiry*, Evidence of Stephen Day, Day 1, p. 169.
9  DTI, *Memorandum*, 2 December 1986, quoted in *Inquiry*, Day 7, p. 188.
10  *Ibid*.
11  FCO, *Summary Record of the IDC*, 8 January 1985, quoted in *Inquiry*, Day 10, p. 73.
12  Prime Minister, *Letter*, 19 August 1986, quoted in *Inquiry*, Day 6, pp. 48–9.
13  MOD, MinDP, *Letter* to Minister of State, FCO, 1984, quoted in *Inquiry*, Day 6, pp. 48–9.
14  *Inquiry*, Evidence of Sir Adam Butler, Day 6, p. 63.
15  Secretary of State for Defence, *Letter* to the Foreign Secretary, 1984, quoted in *Inquiry*, Day 3, p. 93.
16  David Sanders, *Losing an Empire, Finding a Rôle: British Foreign Policy Since 1945*, London, Macmillan, 1990, pp. 186–7.
17  B. R. Pridham (Ed.), *The Arab Gulf and the West*, London, Croom Helm, and University of Exeter, Centre for Arab Gulf Studies, 1985, p. 39.
18  President Carter's 'State of the Union' Address to the Joint Session of the Congress, 23 January 1980.
19  George P. Shultz, *Turmoil and Triumph: My Years as Secretary of State*, New York, Scribners, 1993, p. 237.
20  Donald Rumsfeld, US Envoy to the Middle East, quoted in Shultz, *op cit*, p. 235.
21  Foreign Affairs Committee, *Second Report*, HC 279–I, *op cit*, p. xviii.
22  Memorandum submitted by the FCO, *The Iran–Iraq Conflict*, Foreign Affairs Committee, Minutes of Evidence, 10 February 1988, HC 279–II, p. 27.
23  Foreign Affairs Committee, *Second Report*, HC 279–I, *op cit*, p. xviii.
24  Foreign Affairs Committee Report: *Current UK Policy Towards the Iran–Iraq Conflict*, Foreign and Commonwealth Office Response, Cmd Paper 505, October 1988, p. 3.
25  Margaret Thatcher, *The Downing Street Years*, London, HarperCollins, 1993, p. 162.

26 David Mellor, Minister of State, FCO, Foreign Affairs Committee, *op cit,* Minutes of Evidence, 10 February 1988, HC 279–II, p. 26, Q87.

27 MOD, MinDP, *Letter* to Minister of State, FCO, 4 November 1986, quoted in *Inquiry, op cit,* Day 4, p. 38.

28 *The Observer,* 20 November 1988; and DTI, *Memorandum,* 2 November 1988, quoted in *Inquiry,* Day 41, p. 220.

29 *Hansard,* 16 January 1991, col. 501.

30 DTI, *Memorandum,* 28 March 1990, (000352)*.
(* Numbered documents refer to those released to Paul Henderson and passed to the author.)

31 MOD, *Memorandum,* 15 February 1990, quoted in *Inquiry, op cit,* Day 56, p. 35 and p. 40.

32 FCO, *Minute* of a Meeting between Minister of State and officials, 7 March 1984, quoted in *Inquiry,* Day 1, p. 18.

33 MOD, *Minute,* 13 September 1989, quoted in *Inquiry,* Day 56, p. 25.

34 *Minute* of an inter-departmental meeting, 4 August 1989, quoted in *Inquiry,* Day 17, p. 87.

35 *Inquiry,* Evidence of David Gore-Booth, Day 21, p. 139.

36 *Ibid,* Evidence of Sir Richard Luce, Day 1, pp. 31–2.

37 *Ibid,* Evidence of Sir Stephen Egerton, Day 11, pp. 11–12.

38 DTI, *Brief for Ministerial Meeting,* 21 December 1988, quoted in *Inquiry,* Day 42, p. 11.

39 MOD, *Submission* to the MinDP, 1 March 1989, quoted in *Inquiry,* Day 29, p. 76.

40 FCO, *Summary Record of the IDC,* 27 April 1988, (000060).

41 *The Observer,* 28 February 1988.

42 *Inquiry, op cit,* Evidence of Lt Col. Glazebrook, Day 4, p. 61.

43 Geoffrey Howe, *Conflict of Loyalty,* London, Macmillan, 1994, p. 512.

44 FCO, *Summary Record of the IDC,* 14 March 1989, (000148).

45 *Ibid.*

46 MOD, *Submission* to MinDP, 1 March 1989, quoted in *Inquiry, op cit,* Day 29, pp. 91–2.

47 MOD, Chair of the MODWG, (DESS 2), *Note* of Meeting, 15 June 1989, quoted in *Inquiry,* Day 29, pp. 149–50.

48 FCO, Assistant Under-Secretary for MED and NENAD, *Memorandum,* July 1986, quoted in *Inquiry,* Day 18, pp. 91–2.

49 Private Secretary to the Prime Minister, *Letter* to the Foreign Secretary, 2 September 1988, quoted in *Inquiry,* Day 26, p. 44.

50 Response to a Parliamentary Question, 22 May 1985, quoted in *Inquiry,* Day 1, pp. 110–11; and Minister for Trade, *Letter* to the Prime Minister, 24 November 1986, quoted in *Inquiry,* Day 7, p. 177.

51 Joseph E. Persico, *Casey: From OSS to the CIA,* New York, Penguin, 1991, p. 445.

52 Draft National Security Directive, 'US Policy toward Iran', 11 June 1985, *The Iran-Contra Scandal: The Declassified History,* Peter Kornbluh and Malcolm Byrne (Eds), New York, The New Press, 1993, p. 221.

53 See *Inquiry, op cit:* Evidence of Lord Howe, Day 54, pp. 56–60; William Waldegrave, Day 27, pp. 78–80; Mark Higson, Day 17, p. 91; and Lord Trefgarne, Day 81, p. 162.

54 Private Secretary to the Prime Minister, *Note,* 1 September 1988, quoted in *Report, op cit,* D3.13, p. 376.

55 FCO, *Briefing* for the Ministerial meeting, 21 December 1988, quoted in *Report,* D3.30, p. 384.

56 FCO, *Briefing* for the Foreign Secretary, July 1989, quoted in *Report,* D3.108, p. 419.

57 *Briefing for Ministerial Meeting,* 19 July 1990, the 'Iraq Note', 13 July 1990, quoted in *Inquiry,* Day 27, p. 166.

58 Pridham, *op cit,* p. 39.

59 *Ibid,* p. 89.

60 Frank Brenchley, *Britain and the Middle East: an Economic History 1945–87,* London, Lester Crook, 1989, p. 329.

61 Sanders, *op cit*, p. 119.
62 Brenchley, *op cit*, p. 263.
63 Foreign Affairs Committee, *Second Report*, HC 279–I, *op cit*, pp. xxvi–xxvii.
64 David Mellor, Minister of State, FCO, Foreign Affairs Committee, Minutes of Evidence, 10 February 1988, HC 279–II, p. 25, Q82.
65 Alan Clark, *Diaries*, London, Wiedenfeld and Nicolson, 1993, p. 252.
66 Thatcher, *op cit*, p. 162.
67 J. B. Kelly, *Arabia, the Gulf and the West*, London, HarperCollins, 1980, p. 101.
68 *Ibid*, p. 490.
69 *Ibid*, p. 489.
70 FCO, *Submission*, 9 September 1988, quoted in *Report, op cit*, D3.17, p. 378.
71 Keith Krause, *Arms and the State: Patterns of Military Production and Trade*, Cambridge, Cambridge University Press, 1992, p. 186.
72 Sir Anthony Parsons, Foreign Affairs Committee, Minutes of Evidence, 30 March 1988, HC 279–I, p. 106, Q390.
73 *Ibid*, Q385.
74 The Rt Hon. Alan Clark, *Interview*, 12 April 1995.
75 Kelly, *op cit*, pp. 101 and 103.
76 J. Stanley and M. Pearton, *The International Trade in Arms*, London, Chatto and Windus, 1972, p. 17.
77 MODWG, Lt Col. Glazebrook, Note, 1 August 1988, quoted in *Inquiry, op cit*, Day 4, p. 101.
78 FCO, IDC, *Paper*, 9 September 1988, quoted in *Inquiry*, Day 18, pp. 167–8.
79 FCO, Assistant Under-Secretary for MED and NENAD, *Memorandum*, July 1986, quoted in *Inquiry*, Day 18, pp. 91–2.
80 DTI, *Memorandum*, 7 November 1984, quoted in *Inquiry*, Day 7, pp. 44–5.
81 Supplementary Memorandum submitted by the Foreign and Commonwealth Office, *UK Trade with Gulf States*, Foreign Affairs Committee, Minutes of Evidence, 20 April 1988, HC 279–I, p. 109.
82 *The Daily Telegraph*, 12 March 1979.
83 James A. Bill and Robert Springborg, *Politics in the Middle East*, Fourth Edition, New York, HarperCollins, 1994, p. 433.
84 *The Financial Times*, 5 July 1979.
85 *Inquiry, op cit*, Evidence of Sir Stephen Egerton, Day 11, p. 12.
86 Overseas and Defence Committee of the Cabinet, *Minute*, 29 January 1981, quoted in *Inquiry*, Day 3, p. 30.
87 Sir Anthony Parsons, Foreign Affairs Committee, Minutes of Evidence, 30 March 1988, HC 279–I, p. 99, Q357.
88 Sir John Moberly, Foreign Affairs Committee, Minutes of Evidence, 23 March 1988, HC 279–II, p. 77.
89 FCO, *Memorandum*, 29 October 1984, quoted in *Inquiry, op cit*, Day 10, p. 38.
90 *Report, op cit*, D1.9, p. 154; and D1.12, p. 155.
91 'Exports to Iraq', Memorandum by ECGD, (EQ20), Trade and Industry Committee, *Exports to Iraq*, Appendices to the Minutes of Evidence, 26 November 1991, Session 1991–92, House of Commons Papers 86–i, p. 120.
92 *The Independent*, 10 June 1988.
93 *Report, op cit*, D2.80, pp. 241–2.
94 *Ibid*, D1.12–13, p. 155.
95 *The Independent*, 10 June 1988.
96 *Inquiry, op cit*, Evidence of Sir Richard Luce, Day 1, p. 65.
97 MOD, *Speaking Brief* for MinDP, 17 July 1990, (000408).
98 *The Guardian*, 11 September 1984.
99 *The Independent*, 26 September 1987.
100 Brenchley, *op cit*, pp. 322–24.
101 *Inquiry, op cit*, Evidence of Sir Stephen Egerton, Day 11, p. 43.
102 SIPRI, *The Arms Trade with the Third World*, London, Paul Elek, 1971, p. 42.
103 MOD, *Record of the MODWG Meeting*, 13 December 1984, quoted in *Inquiry, op cit*, Day 4, p. 29; and Evidence of Lt Col. Glazebrook, Day 4, p. 29.

104 *The Independent*, 26 September 1987.

105 *The Financial Times*, 20 April 1988.

106 FCO, Minister of State, *Letter* to Minister for Trade and MinDP, 20 June 1989, quoted in *Inquiry, op cit*, Day 27, pp. 103–4.

107 FCO, *Submission* to the Foreign Secretary, 22 June 1990, quoted in *Inquiry*, Day 57, p. 37.

108 *The Financial Times*, 16 October 1989; and Edward A. Kolodziej, *Making and Marketing Arms: The French Experience and its Implications for the International System*, Princeton, N.J., Princeton University Press, 1987, p. 351.

109 FCO, Assistant Under-Secretary for MED and NENAD, *Memorandum*, 4 July 1986, quoted in *Inquiry, op cit*, Day 18, p. 83.

110 Foreign Secretary, *Letter* to the Prime Minister, 1 December 1986, quoted in *Inquiry*, Day 18, p. 98.

111 FCO, *Briefing* for Minister of State, 20 December 1988, quoted in *Inquiry*, Day 18, p. 130.

112 David Mellor, Minister of State, FCO, Foreign Affairs Committee, Minutes of Evidence, 10 February 1988, HC 279–II, p. 26, Q87.

113 Sanders, *op cit*, p. 291.

114 FCO, IDC, *Paper*, 9 September 1988, quoted in *Inquiry, op cit*, Day 18, p. 167.

115 MOD, *Memorandum*, 9 March 1990, quoted in *Inquiry*, Day 56, p. 40.

116 *Inquiry*, Evidence of David Mellor, MP, Day 25, p. 55.

117 John Greenaway, 'Models of British Politics', in John Greenaway, Steve Smith and John Street, *Deciding Factors in British Politics: A Case Studies Approach*, London, Routledge, 1992, p. 66.

118 FCO, Minister of State, *Memorandum*, 20 June 1989, quoted in *Inquiry, op cit*, Day 27, p. 104.

119 Trade and Industry Committee, *Second Report, Exports to Iraq: Project Babylon and Long-Range Guns*, Session 1991–92, House of Commons Papers 86, p. xi.

## Chapter 5

1 Michael Klare, 'Secret Operatives, Clandestine Trades: The Thriving Black Market for Weapons', *Bulletin of the Atomic Scientists*, April 1988, pp. 16–24.

2 Aaron Karp, 'The Rise of Black and Grey Markets', *The Annals of the American Academy of Political and Social Science*, Robert E. Harkavy and Stephanie G. Neuman (Eds), *The Arms Trade: Problems and Prospects in the Post-Cold War World*, Thousand Oaks, Sage, 1994, p. 176.

3 Klare, *op cit*, p. 19.

4 Karp, *op cit*, p. 175.

5 *The Times*, 7 April 1984.

6 See: *The Daily Telegraph*, 24 April 1986; and *The Sunday Telegraph*, 14 September 1986.

7 AWP, *Minute*, 15 July 1986, quoted in *Inquiry into Exports of Defence Equipment and Dual-Use Goods to Iraq*, Hearings in the presence of the Rt Hon. Lord Justice Scott, Day 12, p. 84.

8 Chris Cowley, *Guns, Lies and Spies*, London, Hamish Hamilton, 1992; and Sir Hal Miller, *Interview*, 12 July 1994.

9 Trade and Industry Committee, *Second Report, Exports to Iraq: Project Babylon and Long-Range Guns*, Session 1991–92, House of Commons Papers 86, p. xxxii.

10 Security Service, *Report*, 23 July 1981, quoted in the Rt Hon. Sir Richard Scott, the Vice-Chancellor, *Report of the Inquiry into the Export of Defence and Dual-Use Goods to Iraq and Related Prosecutions*, House of Commons Papers 115, 1996, E6.1, p. 869.

11 *Jane's Defence Weekly*, 8 September 1984.

12 JIC, *Paper*, 2 December 1985, quoted in *Report, op cit*, E4.1, p. 861.

13 MOD, DIS, *Loose Minute*, 29 November 1984, quoted in *Inquiry, op cit*, Day 12, pp. 2–3.

14 SIS, *Report*, 1985, quoted in *Report, op cit*, E3.1, pp. 853–4.

15 Private Secretary to the Prime Minister, *Minute* to the Prime Minister, 29 January 1986, quoted in *Inquiry, op cit*, Day 24, p. 123.

16 MOD, Lt Col. Glazebrook, *Memorandum* to DESS 2, 26 November 1987, quoted in *Inquiry*, Day 12, pp. 5–7; and FCO, *Summary Record of the IDC*, 6 January 1988, quoted in *Inquiry*, Day 14, p. 62.

17 *Inquiry*, Day 11, p. 152.

18 *Hansard*, 2 December 1980, col. 155.

19 FCO, *Memorandum*, 1 October 1983, quoted in *Inquiry, op cit*, Day 11, p. 128.

20 See: *Report, op cit*, E2.4–5, pp. 820–1.

21 MOD, *Paper*, 1984, quoted in *Report*, E2.8, pp. 821–2.

22 MOD, *Letter* to MED, FCO, 30 March 1984, quoted in *Report*, D2.361, p. 337.

23 *Inquiry, op cit*, Day 22, pp. 22–3.

24 JIC, *Report*, December 1985, quoted in *Inquiry*, Day 31, p. 6.

25 SIS, *Reports*, 14 March 1986 and 12 November 1986, quoted in *Inquiry*, Day 24, p. 106 and pp. 110–111. See also reports quoted in *Inquiry*, Day 31, pp. 118–22.

26 See: *Report, op cit*, E2.17, pp. 828–33.

27 British Embassy, Amman, *Letter* to FCO, 21 December 1987, quoted in *Report*, E2.14, p. 824.

28 *Inquiry, op cit*, Evidence of David Mellor MP, Day 25, p. 131.

29 *Ibid*, Evidence of Sir Stephen Egerton, Day 11, pp. 115–16.

30 Quoted in 'Special Assignment', BBC Radio 4, 30 April 1993.

31 *Inquiry, op cit*, Evidence of Simon Fuller, Day 31, pp. 67–8.

32 See, for example: James Edmiston, *The Sterling Years*, Barnsley, Leo Cooper, 1995, p. 132.

33 MOD, Note on an AWP Application, 13 March 1985, quoted in *Inquiry, op cit*, Day 12, pp. 26–8.

34 MOD, Comments on an AWP Application, 27 May 1986, quoted in *Inquiry*, Day 12, p. 78.

35 MOD, Comments on an ELA, 1 October 1986, quoted in *Inquiry*, Day 12, pp. 100–102.

36 FCO, *Summary Record of the IDC*, 11 August 1988, quoted in *Inquiry*, Day 19, pp. 23–4; and Day 13, pp. 10–12 and pp. 16–18.

37 MODWG, *Minute* to Defence Sales, 1 July 1985, quoted in *Inquiry*, Day 6, pp. 159–60.

38 *Inquiry*, Evidence of Sir Stephen Egerton, Day 11, p. 101.

39 *Ibid*, Evidence of Simon Fuller, FCO, Day 31, p. 23.

40 *Ibid*, Evidence of Lt Col. Glazebrook, Day 12, pp. 130–1.

41 *Report, op cit*, E2.58, p. 849.

42 *Inquiry, op cit*, Evidence of Sir Adam Butler, Day 6, p. 164.

43 *Ibid*, Day 12, pp. 130–1.

44 *Report, op cit*, E2.13, p. 823.

45 MOD, DIS, *Minute* to DESS 2 and the IDC, 8 May 1990, quoted in *Inquiry, op cit*, Day 13, pp. 82–3.

46 *Inquiry*, Evidence of Lt Col. Glazebrook, Day 12, p. 172 and p. 131.

47 MOD, Lt Col. Glazebrook, *Memorandum*, 18 February 1987, quoted in *Inquiry*, Day 12, p. 133.

48 *Inquiry*, Day 12, pp. 134–5.

49 *Report, op cit*, E8.3, p. 892.

50 *Ibid*, E2.62, p. 851.

51 *Inquiry, op cit*, Evidence of Rob Young, Day 23, pp. 128–9.

52 *Ibid*, Evidence of Mark Higson, Day 17, p. 105.

53 *Ibid*, Evidence of Alan Clark, Day 50, pp. 25–6.

54 Svenska Freds-och Skiljedomsforeningen, (The Swedish Peace and Arbitration Society, SPAS), *The International Connections of the Bofors Affair*, Stockholm, SPAS, December 1987.

55 Cartel documents passed to the author.

56 Quoted in *Private Eye*, 16 October 1987.

57 Cartel documents passed to the author.
58 Quoted in *Private Eye*, 2 October 1987.
59 SPAS, *op cit*, p. 20.
60 *Report, op cit*, D7.14, pp. 782–3.
61 *Ibid*, D7.12, p. 782.
62 Quoted in *The Guardian*, 30 December 1993.
63 Roger Freeman, Minister for Defence Procurement, House of Commons Written Answer to Llew Smith, MP, quoted in *The Independent*, 7 July 1995.
64 *Report, op cit*, D7.4, p. 778.
65 *Ibid*, E7.15, p. 885.
66 *Hansard*, 23 November 1992, col. 696.
67 *The Independent on Sunday*, 22 November 1992.
68 *Ibid*.
69 SIS, *Report*, April 1987, quoted in *Report, op cit*, E7.5, p. 881.
70 Customs Export Licensing Policy Branch, John Fisher, Customs ID, *Report*, September 1988, quoted in *Report*, E7.14, p. 885.
71 MOD, Lt Col. Glazebrook, *Minute*, 6 June 1988, quoted in *Report*, E7.4, p. 879.
72 See: *Report*, E10.12, p. 914.
73 *The Independent on Sunday*, 30 August 1992.
74 *The Independent*, 7 July 1995, quoting documents in its possession.
75 *Report, op cit*, E3.13, p. 859; and E6.6, p. 871.
76 *Ibid*, E6.10, p. 873.
77 *Minutes of the REU*, 15 February 1988, quoted in *Inquiry, op cit*, Day 11, p. 139.
78 Lord Younger, quoted in *The Independent*, 31 August 1992.
79 *Report, op cit*, D7.59, p. 796.
80 *The Guardian*, 14 June 1995.
81 *Jane's Defence Weekly*, 3 October 1987.
82 Memorandum supplied by the Department of Trade and Industry to the Trade and Industry Committee, *Export Licensing and BMARC*, Minutes of Evidence, 6 December 1995, Session 1995–96, House of Commons Papers, 87–i, p. 7.
83 Quoted in *The Guardian*, 15 June 1995.
84 *The Observer*, 1 April 1984.
85 Quoted in *The Guardian*, 20 June 1995.
86 *Report, op cit*, D7.81, p. 804; and D7.91, p. 807.
87 *The Guardian*, 15 June 1995.
88 *Ibid*.
89 See: Trade and Industry Committee, HC 87–ii and iii, *op cit*; and *Report, op cit*, D7.77, p. 803.
90 Gerald James, quoted in *The Guardian*, 20 June 1995.
91 The Rt Hon. Alan Clark, *Interview*, 12 April 1995.
92 Quoted in *Time*, 2 September 1991.
93 Henry B. Gonzalez, Chairman Committee on Banking, Finance and Urban Affairs, *Statement Update on BNL Investigation*, 21 January 1993.
94 *Order of the US District Court*, Northern District of Georgia, Atlanta Division, United States of America vs DeCardis *et al*, Marvin N. Shoob, Senior Judge, 23 August 1993.
95 *The Daily Telegraph*, 18 December 1987.
96 *The Financial Times*, 24 September 1987.
97 *Los Angeles Times*, 3 September 1985; and *The Sunday Times*, 18 August 1985.
98 *The Guardian*, 24 September 1987.
99 *South*, February 1987.
100 FCO, *Summary Record of the IDC*, September 1986, quoted in *Inquiry, op cit*, Day 19, pp. 2–3.
101 *Inquiry*, Evidence of Sir David Miers, Day 19, pp. 5–6.
102 *Ibid*, Evidence of Lt Col. Glazebrook, Day 12, p. 102 and pp. 7–8.
103 *Ibid*, Evidence of Sir David Miers, Day 19, pp. 29–30.
104 *The Daily Telegraph*, 2 January 1988.
105 Quoted in *The Guardian*, 22 February 1984.

106 FCO, *Summary Record of the IDC*, 27 April 1988, (000060)*.
(* Numbered documents refer to those released to Paul Henderson and passed to the author.)

107 FCO, *Summary Record of the IDC*, 13 December 1989, (000340).

108 *Jane's Defence Weekly*, 13 December 1986.

109 FCO, *Summary Record of the IDC*, 13 December 1989 (000339); and *Summary Record of the IDC*, 23 October 1987, quoted in *Inquiry, op cit*, Day 14, p. 39.

110 *Inquiry*, Evidence of Lt Col. Glazebrook. Day 4, pp. 22–3.

111 See for example: *Inquiry*, Evidence of David Gore-Booth, Day 22, p. 28; and Evidence of Simon Fuller, Day 31, p. 92.

112 *Ibid*, Evidence of Lady Thatcher, Day 48, pp. 222–4; Evidence of Simon Fuller, Day 31, p. 65; and *Report, op cit*, E2.20, p. 834.

113 *Inquiry*, Evidence of Simon Fuller, Day 31, p. 208.

114 *Ibid*, Evidence of Charles Haswell, Day 33, p. 11; and Evidence of Timothy Renton, MP, Day 24, p. 106.

115 *Ibid*, Evidence of Sir David Miers, Day 19, p. 29.

116 *Ibid*, pp. 5–6.

117 *Minutes of the REU*, 25 November 1988, quoted in *Inquiry*, Day 11, p. 140.

118 John Reed, Editor, *Defence Industries' Digest*, quoted in 'Special Assignment', BBC Radio 4, 30 April 1993.

119 R. Anderson, DSO, Expenditure Committee, Minutes of Evidence, 4 November 1975, Session 1975–76, House of Commons Papers, 155–I, p. 215, Q804.

120 *Inquiry, op cit*, Evidence of Paul Channon, Day 7, pp. 24–5.

121 *Ibid*, Evidence of Lord Howe, Day 54, p. 4.

122 R. Aliboni, 'Recent Developments in the Gulf Crisis: The West and the Gulf', *Journal of Arab Affairs*, Vol. 7, No,1, Spring 1988, p. 58.

123 *Birmingham Daily News*, 6 August 1987.

124 Paul Henderson, *Interview*, 22 September 1993.

125 SIS, *Telegrams*, 6 January 1989, (000113–4).

126 Kenneth Timmerman, *The Nation*, July 18/25, 1987, p. 47.

127 Anthony H. Cordesman, 'After the Gulf War: The World Arms Trade and Its Arms Races in the 1990s', *Brassey's Defence Yearbook*, 1992, pp. 221–2.

128 The Secretary of State for Defence, *Hansard*, 23 November 1992, col. 704.

129 Anthony H. Cordesman, 'Arms to Iran: The Impact of US and Other Sales on the Iran–Iraq War', *American-Arab Affairs*, Vol. 20, Spring 1987, p. 29.

130 Quoted in Mohamed Heikal, *Illusions of Triumph: An Arab View of the Gulf War*, London, HarperCollins, 1993, p. 81.

131 *Ibid*.

132 George P. Shultz, *Turmoil and Triumph: My Years as Secretary of State*, New York, Scribners, 1993, p. 236.

133 Charles Tripp, 'Europe and the Persian Gulf', *Cambridge Review of International Affairs*, Vol. 2, No. 2, 1988, p. 1.

134 Lord Carrington, Foreign Secretary, Foreign Affairs Committee, Minutes of Evidence, 3 December 1980, Session 1980–81, House of Commons Papers 41–i, Q30.

135 Foreign Affairs Committee, *Second report, Current UK Policy towards the Iran/Iraq Conflict*, Session 1987–88, House of Commons Papers, 279–I, p. xxiii.

136 Sir Anthony Parsons, Foreign Affairs Committee, Minutes of Evidence, 30 March 1988, HC 279–II, pp. 102–3, Q371.

137 Margaret Thatcher, *The Downing Street Years*, London, HarperCollins, 1993, p. 91.

138 *Ibid*, pp. 163–4.

139 John Goulden, Assistant Under-Secretary of State, FCO, Trade and Industry Committee, *op cit*, Minutes of Evidence, 28 January 1992, HC 86–vi, p. 287, Q2020.

140 *Inquiry, op cit*, Evidence of Mr Blackley, Day 20, p. 46.

141 Edgar O'Ballance, *The Gulf War*, London, Brassey's, 1988, pp. 207–8.

142 Foreign Affairs Committee, *Second Report, op cit*, HC 279–I, p. vii.

143 O'Ballance, *op cit*, p. 207.

144 Thomas Ohlson and Michael Brzoska, 'SIPRI Report', in *Ploughshares Monitor*, June 1984.

145 The Rt Hon. Alan Clark, *Interview*, 12 April 1995.
146 For accounts of the Iran–Iraq War, see: Jasim M. Abdulghani, *Iran and Iraq: The Years of Crisis*, London, Croom Helm, 1984; Shahram Chubin and Charles Tripp, *Iran and Iraq at War*, London, I. B. Tauris, 1988; Anthony H. Cordesman, *The Iran–Iraq War and Western Security 1984–87: Strategic Implications and Policy Options*, London, Jane's Publishing Company, 1987; Dilip Hiro, *The Longest War: The Iran–Iraq Military Conflict*, London, Paladin Books, 1990; Efram Karsh, *The Iran–Iraq War: A Military Analysis*, Adelphi Papers, London, International Institute for Strategic Studies, 1987; Majid Khadduri, *The Gulf War: The Origins and Implications of the Iran–Iraq Conflict*, Oxford, Oxford University Press, 1988; Edgar O'Ballance, *The Gulf War*, London, Brassey's, 1988; and Shirin Tahir-Kheli and Shaheen Ayubi (Eds), *The Iran–Iraq War: New Weapons, Old Conflicts*, New York, Praeger, 1983.
147 *Memorandum* by Sir John Moberly, Foreign Affairs Committee, Minutes of Evidence, 23 March 1988, HC 279–II, p. 74.
148 Quoted in Chaim Herzog, 'A Military-Strategic Overview', in Efraim Karsh, (Ed.), *The Iran–Iraq War: Impact and Implications*, London, Macmillan Press, 1989, p. 255.
149 SIPRI, *The Arms Trade with the Third World*, London, Paul Elek, 1971, p. 41.
150 William S. Olson, 'The Iran–Iraq War: A Dialogue of Violence', *Defence Analysis*, Vol. 2, No. 3, 1986, p. 236.
151 Cordesman, 'Arms to Iran', *op cit*, p. 19.
152 British Embassy, Amman, *Telex* to FCO, 11 October 1986, quoted in *Inquiry, op cit*, Day 24, p. 105.
153 Anthony H. Cordesman, *After the Storm: The Changing Military Balance in the Middle East*, London, Mansell Publishing Ltd, 1993, p. 384.
154 Gerald James, *Letter* to Sir Richard Scott, 22 February 1995. See also: Gerald James, *In the Public Interest*, London, Little, Brown & Company, 1995.
155 Gerald James, *Letter, op cit*.
156 *Ibid*; and Paul Henderson, *The Unlikely Spy*, London, Bloomsbury, 1993, p. 80.
157 *Report, op cit*, D5.41, p. 542.
158 Quoted in Joseph E. Persico, *Casey: From OSS to the CIA*, New York, Penguin, 1991, p. 369.
159 Shultz, *op cit*, p. 813.
160 Robert E. Harkavy, 'Arms Resupply During Conflict: A Framework for Analysis', *The Jerusalem Journal of International Relations*, Vol. 7, No. 3, 1985, p. 37.
161 The Rt Hon. Alan Clark, *Interview*, 12 April 1995.
162 J. Stanley and M. Pearton, *The International Trade in Arms*, London, Chatto and Windus, 1972, p. 17.
163 The Rt Hon. Alan Clark, *Interview*, 12 April 1995; and Gerald James, *Letter* to Sir Richard Scott, 22 February 1995.
164 Shultz, *op cit*, p. 926.
165 Alan Clark, *Diaries*, London, Wiedenfeld and Nicolson, 1993, p. 321.
166 Geoffrey Howe, *Conflict of Loyalty*, London, Macmillan, 1994, p. 545.
167 David Sanders, *Losing an Empire, Finding a Rôle: British Foreign Policy Since 1945*, London, Macmillan, 1990, p. 185.
168 Sir Anthony Parsons, Foreign Affairs Committee, Minutes of Evidence, 30 March 1988, HC 279–II, p. 105, Q381.
169 Yezid Sayigh, *Arab Military Industry: Capability, Performance and Impact*, London, Brassey's, 1992, p. 105.
170 Kenneth R. Timmerman, *The Death Lobby: How the West Armed Iraq*, London, Fourth Estate, 1992, p. 227.
171 Directorate of Intelligence, Central Intelligence Agency, *Iraq: No End in Sight to Debt Burden* (Summary), 12 April 1990.
172 Heikal, *op cit*, p. 240.
173 Hiro, *op cit*, p. 230.
174 *Inquiry, op cit*, Evidence of Michael Petter, Day 58, p. 35.
175 *Report, op cit*, D2.97, p. 247.
176 *Inquiry, op cit*, Evidence of Sir Stephen Egerton, Day 11, pp. 12–13.

177 *Ibid.*
178 *Ibid*, Evidence of Timothy Renton, MP, Day 24, p. 60.
179 MOD, MinDP, *Letter* to the Minister of State, FCO, November 1984, quoted in *Inquiry*, Day 6, p. 18.
180 Heikal, *op cit*, p. 205.
181 O'Ballance, *op cit*, p. 208.

Chapter 6
1   Michael Klare, 'Secret Operatives, Clandestine Trades: The Thriving Black Market for Weapons', *Bulletin of the Atomic Scientists*, April 1988, p. 19.
2   Michael Klare, 'The Unnoticed Arms Trade: Exports of Conventional Arms-Making Technology', *International Security*, Vol. 8, No. 2, Fall 1983, p. 70.
3   J. Fred Bucy, quoted in *Ibid.*
4   *Inquiry into Exports of Defence Equipment and Dual-Use Goods to Iraq*, Hearings in the presence of the Rt Hon. Lord Justice Scott, Evidence of Lt Col. Glazebrook, Day 13, p. 43.
5   Andrew L. Ross, 'Do-It-Yourself Weaponry', *Bulletin of Atomic Scientists*, May 1990, p. 22.
6   International Atomic Energy Agency, *Foreign Supplies to the Iraqi Nuclear Programme*, Press Release, 11 December 1991.
7   Quoted by Alan George, *Evening Standard*, 8 February 1993.
8   *The Engineer*, 26 May 1994.
9   Secretary of State for Trade and Industry, *Hansard*, 15 April 1991, col. 30.
10  *The Financial Times*, 1 May 1991.
11  *The Guardian*, 17 October 1991.
12  MOD, *British Assistance to the Emerging Iraqi Arms Industry*, MODWG/DIS Survey, 29 October 1990, (000428)* .
    (* Numbered documents refer to those released to Paul Henderson and passed to the author.)
13  See: the Rt Hon. Sir Richard Scott, the Vice-Chancellor, *Report of the Inquiry into the Export of Defence Equipment and Dual-Use Goods to Iraq and Related Prosecutions*, House of Commons Papers 115, 1996, D1.98–9 pp. 185–6; and D2.62, p. 236, D6.55–72, pp. 587–94 and D6.194–220, pp. 652–67.
14  Kenneth Timmerman, quoted in 'Today', BBC Radio Four, 22 March 1993.
15  *British Assistance to the Emerging Iraqi Arms Industry, op cit*, (000429–30).
16  *Ibid.*
17  *Middle East*, December 1989, p. 29.
18  MOD, *British Assistance to the Emerging Iraqi Arms Industry, op cit*, (000429).
19  *Hansard*, 27 November 1992, col. 864.
20  *Supplementary Memorandum: Export Controls in Relation to Iraq*, by the Department of Trade and Industry, (EQ18), 1 November 1991, Annex C, Table 2: Breakdown of Exports of Chemicals by Shipments, Trade and Industry Committee, *Exports to Iraq*, Minutes of Evidence, 26 November 1991, Session 1991–92, House of Commons Papers, 86–i, pp. 108–9.
21  *Letter to the Clerk of the Committee from Dr Alastair Hay (EQ22): Export Controls in Relation to Iraq*, 12 November 1991, Trade and Industry Committee, Minutes of Evidence, 26 November 1991, HC 86–i, p. 121.
22  *Supplementary Memorandum: Export Controls in Relation to Iraq, op cit*, Trade and Industry Committee, HC 86–i, p. 109.
23  United Nations Security Council, *Report of the Specialist Appointed by the Secretary-General to Investigate Allegations by the Islamic Republic of Iran Concerning the Use of Chemical Weapons*, S/16433, 26 March 1984, pp. 11–12, quoted in Edward M. Spiers, *Chemical and Biological Weapons: A Study of Proliferation*, London, Macmillan, 1994, p. 18.
24  *Letter to the Clerk of the Committee from Dr Alastair Hay, (EQ22): Export Controls in Relation to Iraq, op cit*, Trade and Industry Committee, HC 86–i, p. 122.
25  Private Secretary to the Prime Minister, *Memorandum*, 29 January 1986, quoted in *Inquiry, op cit*, Day 24, p. 123.

26  *Inquiry*, Evidence of Timothy Renton, MP, Day 24, pp. 124–6.

27  FCO, *Telex* to the Embassy Cairo, Autumn 1986, quoted in *Inquiry*, Day 24, p. 134.

28  MOD, DIS, *Loose Minute*, 29 November 1984, quoted in *Inquiry*, Day 12, pp. 2–3.

29  *Hansard*, 16 January 1991, col. 501.

30  *Hansard*, 31 January 1991, col. 560.

31  *Briefing for Ministerial Meeting*, 19 July 1990, the 'Iraq Note', 13 July 1990, quoted in *Report, op cit*, D3.155, p. 441.

32  *Inquiry, op cit*, Evidence of Lady Thatcher, Day 48, pp. 60–1.

33  Dr John Hassard, quoted in 'Today', BBC Radio Four, 22 March 1993.

34  Yezid Sayigh, *Arab Military Industry: Capability, Performance and Impact*, London, Brassey's, 1992, p. 103.

35  Kenneth R. Timmerman, *The Death Lobby: How the West Armed Iraq*, London, Fourth Estate, 1992, pp. 48–9.

36  *Ibid*, p. 352.

37  *Middle East News*, 8 May 1989.

38  *Hansard*, 30 April 1990, col. 405.

39  *Memorandum* submitted by the DTI, (EQ6), Trade and Industry Committee, *Exports to Iraq*, Memoranda of Evidence, 17 July 1991, Session 1990–91, House of Commons Papers 607, pp. 17–18.

40  *The Washington Post*, 15 February 1993.

41  FCO, *Memorandum*, 29 June 1988, quoted in *Inquiry, op cit*, Day 19, p. 85.

42  Foreign Secretary, *Letter* to the Home Secretary, June 1988, quoted in *Inquiry*, Day 19, pp. 103–4.

43  *Inquiry*, Evidence of Lt Col. Glazebrook, Day 4, pp. 144–5.

44  *Ibid*, Evidence of Alan Barrett, Day 39, p. 35.

45  *Ibid*, Evidence of Lt Col. Glazebrook, Day 4, pp. 150–1.

46  *Ibid*, Evidence of Alan Barrett, Day 38, p. 115.

47  Quoted in *Report, op cit*, D5.25, p. 524.

48  FCO, *Matrix Churchill: Export of Lathe Equipment to Iraq*, 1 February 1989, (000131).

49  *Inquiry, op cit*, Evidence of Rob Young, Day 23, p. 56.

50  *Inquiry*, Day 41, pp. 175–6.

51  *Minute* of the WGIP Meeting, 16 May 1989, quoted in *Inquiry*, Day 41, p. 177.

52  FCO, Minister of State, Comment, 8 August 1989, on a *Submission*, 4 August 1989, quoted in *Inquiry*, Day 26, pp. 15–16.

53  SIS, *Reports*, 5 September 1989, quoted in *Inquiry*, Day 47, p. 31; and GCHQ *Report*, 12 September 1989, quoted in *Inquiry*, Day 22, p. 83.

54  *The Times*, 9 November 1989, quoted in Lt Col. Glazebrook, *Memorandum* to the Assistant Chief of the Defence Staff, Operational Requirements (Land) November 1989, quoted in *Inquiry*, Day 5, p. 3.

55  FCO, *Summary Record of the IDC*, 13 December 1989, sent to the Minister of State, 18 December 1989 (000343).

56  *The Guardian*, 8 February 1990; and *Jane's Defence Weekly*, 21 July 1990.

57  Quoted in *Report, op cit*, D5.25, p. 530.

## Chapter 7

1  *Inquiry into Exports of Defence Equipment and Dual-Use Goods to Iraq*, Hearings in the presence of the Rt Hon. Lord Justice Scott, Evidence of Eric Beston, Day 52, p. 89.

2  The Secretary of State for Trade and Industry, *Hansard*, 18 April 1990, col. 1427.

3  The Secretary of State for Trade and Industry, *Hansard*, 13 December 1990, col. 493; and 18 April 1990, cols. 1433 and 1429.

4  Trade and Industry Committee, *Second Report, Exports to Iraq: Project Babylon and Long-Range Guns*, Session 1991–92, House of Commons Papers 86, p. xliii.

5  The Secretary of State for Trade and Industry, *Hansard*, 18 April 1990, col. 1427; *Hansard*, 18 December 1990, col. 100; and *Hansard*, 18 April 1990, col. 1427.

6  Minister of State for Defence Procurement, *Hansard*, 25 April 1990, col. 210.

7  The Secretary of State for Trade and Industry, *Hansard*, 18 April 1990, col. 1427.

8   Trade and Industry Committee, *Second Report, op cit*, HC 86, p. xv.
9   Notes seen by the writer. Sir Hal Miller, *Interview*, 12 July 1994.
10  *Hansard*, 18 April 1990, col. 1432.
11  *Inquiry, op cit*, Evidence of Sir Hal Miller, Day 8, p. 2.
12  *Ibid*, pp. 9–10.
13  The Rt Hon. Sir Richard Scott, the Vice-Chancellor, *Report of the Inquiry into the Export of Defence Equipment and Dual-Use Goods to Iraq and Related Prosecutions*, House of Commons Papers 115, 1996, F2.24, p. 959.
14  SIS, Mr L, *Manuscript Notes*, 7 July 1988, quoted in *Report*, F2.13, pp. 952–3.
15  Quoted in *Report*, F2.25, p. 960.
16  *Report*, F2.26, p. 963.
17  *Ibid*, F2.28, pp. 964–5 and F2.41–2, pp. 971–2.
18  Quoted in *Ibid*, F2.42, p. 973.
19  *Ibid*, F2.46, pp. 974–5 and F2.84, p. 993.
20  *Ibid*, F2.47–8, pp. 975–6.
21  *Report*, F2.54, p. 978, F2.61, p. 981 and F2.71, p. 986.
22  Quoted in *Ibid*, E2.73, p. 987.
23  *Ibid*, E2.76, pp. 987–8 and E2.78, p. 989.
24  MOD, *Letter* to the Cabinet Office, 24 April 1990, quoted in *Report*, F2.79, p. 990.
25  *Report*, F2.81 and F2.83, p. 992.
26  Security Service, *Loose Minute*, 4 November 1988, quoted in *Report*, F2.88, p. 994.
27  *Report*, F2.92, p. 996 and F2.93–4, p. 997.
28  *Ibid*, F2.96, p. 998.
29  *Ibid*, F3.1–4, pp. 1003–4.
30  Quoted in *Ibid*, F3.5, p. 1005.
31  Minister for Defence Procurement, Trade and Industry Committee, *Exports to Iraq*, Minutes of Evidence, 27 February 1992, Session 1991–92, House of Commons Papers 86–xv, p. 490, Q3824.
32  *Inquiry, op cit*, Evidence of Sir Hal Miller, Day 8, pp. 15–16.
33  *Report, op cit*, F3.92, p. 1052.
34  *Ibid*, F3.13–16, pp. 1007–8.
35  Trade and Industry Committee, *Second Report, Exports to Iraq: Project Babylon and Long-Range Guns*, Session 1991–92, House of Commons Papers 86, p. xxxiv.
36  *Report, op cit*, F3.30, p. 1015.
37  *Ibid*, F3.31–6, pp. 1016–19.
38  SIS, *Briefing Note*, 6 October 1989, quoted in *Report*, F3.37, p. 1020.
39  *Report*, F3.38, p. 1021, F3.55, p. 1029, F3.58, p. 1031 and F3.39, p. 1021.
40  *Ibid*, F3.87, pp. 1048–9.
41  See: *Ibid*, F4.1–20, pp. 1055–63.
42  *Ibid*, F4.26–43, pp. 1065–74 and F4.54–79, pp. 1079–94.
43  *The Independent*, 24 February 1992.
44  *Ibid*.
45  *Report, op cit*, J1.42, p. 1621.
46  *The Guardian*, 31 May 1990.
47  *Inquiry, op cit*, Evidence of Sir Hal Miller, Day 8, p. 59.
48  *The Guardian*, 16 January 1992.
49  Dr Chris Cowley, *Letter* to Andrew Kennon, Clerk to the Trade and Industry Committee, 19 February 1992, quoted in Mark Phythian, 'Britain and the Supergun', *Crime, Law and Social Change*, Vol. 19, 1993, p. 370.
50  *Ibid*, p. 376.
51  William Lowther, *Arms and the Man: Dr Gerald Bull, Iraq and the Supergun*, London, Macmillan, 1991, p. 238.
52  *Report, op cit*, F2.6–7, pp. 949–50.
53  Phythian, *op cit*, p. 370. Wong is mentioned with the pseudonym Michael Chang in Lowther, *op cit*, p. 124.
54  *Report, op cit*, F2.5, pp. 948–9 and F2.8, p. 950.
55  *The Independent*, 19 November 1992.

56  FCO, Trade and Industry Committee, Minutes of Evidence, 28 January 1992, HC 86–vii, p. 291, Q2070.
57  Sir Hal Miller, *Interview*, 12 July 1994.
58  *Ibid.*
59  *Report, op cit*, F2.35, p. 969.
60  *Inquiry, op cit*, Evidence of Sir Hal Miller, Day 8, p. 1.
61  *Report, op cit*, F2.36, p. 969.
62  *Inquiry, op cit*, Evidence of Eric Beston, Day 41, p. 179.
63  Treasury Counsel, *Opinion*, 29 October 1990, quoted in *Report, op cit*, J1.21, p. 1606.
64  *Report*, F2.78, p. 989.
65  Treasury Counsel, *Opinion*, 29 October 1990, quoted in *Report*, J1.21, p. 1606.
66  *Ibid.*
67  Quoted in *Report*, D5.4, p. 512.
68  MOD, DIS, Mr JJ, *Note* to Mr Weir, 14 June 1988, quoted in *Report*, F2.22, pp. 958–9.
69  Treasury Counsel, *Opinion*, 24 July 1990, quoted in *Inquiry, op cit*, Day 9, p. 57.
70  *Report, op cit*, F2.9, p. 950.
71  *Ibid*, F2.22, p. 958.
72  Quoted in *Report*, D5.4, p. 512.
73  Quoted in *Ibid*, F2.11, p. 951.
74  See Trade and Industry Committee, Minutes of Evidence, 21 January 1992, HC 86–v, p. 251, Q1582.
75  Quoted in *Report, op cit*, F2.95, p. 997.
76  Quoted in *Ibid*, F4.80, p. 1095.
77  Quoted in *Ibid*, D5.39, p. 541.
78  Trade and Industry Committee, *Second Report, op cit*, HC 86, p. xv.
79  MOD, *British Assistance to the Emerging Iraqi Arms Industry*, MODWG/DIS Survey, 29 October 1990, (000434)*.
    (* Numbered documents refer to documents released to Paul Henderson and passed to the author.)
80  Gerald James, Trade and Industry Committee, Minutes of Evidence, 5 February 1992, HC 86–x, p. 359, Q2404.
81  Trade and Industry Committee, *Second Report, op cit*, HC 86, p. xv.
82  *Report, op cit*, F3.89, p. 1049.
83  Doug Hoyle, MP, quoted in *The Independent*, 29 February 1992.

### Chapter 8

1   Head of DESS, covering note for *Minute* to MinDP, 18 January 1988, quoted in *Inquiry into Exports of Defence Equipment and Dual-Use Goods to Iraq*, Hearings in the presence of the Rt Hon. Lord Justice Scott, Day 43, p. 63.
2   Embassy, Baghdad, *Loose Telex* to MOD, 14 April 1988, quoted in *Inquiry*, Day 22, pp. 47–8.
3   SIS, *Telegram*, 6 January 1989, (000113)*.
    (* Numbered documents refer to those released to Paul Henderson and passed to the author.)
4   FCO, *Matrix Churchill: Export of Lathe Equipment to Iraq*, 1 February 1989, (000132).
5   MOD, DIS, *Loose Minute*, 23 December 1988, quoted in *Inquiry, op cit*, Day 30, p. 7.
6   MOD, DIS, *Loose Minute*, 11 January 1989, (000117).
7   *The Mail on Sunday*, 23 June 1984.
8   ECGD, *Matrix Churchill Ltd, Industrias Cardoen Ltda – Chile – C167/168*, 20 January 1989, (000119).
9   Royal Ordnance Factory, *Hazard Assessment, Cardoen Plant, Chile*, 1987, report passed to the writer.
10  GCHQ, *Chile–Iraq Military Trade: Chilean Arms Firm to Turn Over Munitions Factory to Iraqi Government*, 13 October 1989, quoted in *Inquiry, op cit*, Day 30, pp. 73–4.
11  *Minute of the REU Meeting*, 8 December 1989, quoted in *Inquiry*, Day 47, p. 131.
12  *Inquiry*, Evidence of Anthony Steadman, Day 47, pp. 122–3.

13  See for example, *Inquiry*, Evidence of Eric Beston, Day 44, p. 61.

14  Minute of the WGIP Meeting, 23 June 1989, (000321).

15  SIS, *Reports*, 12 July 1989; 5 September 1989; 12 September 1989, quoted in *Inquiry*, *op cit*, Day 47, pp. 31–2.

16  SIS, *Report*, 12 September 1989, quoted in *Inquiry*, Day 22, p. 83.

17  SIS, *Report*, 5 September 1989, quoted in *Inquiry*, Day 47, p. 33.

18  Matrix Churchill, *Letter* to the DTI, 13 July 1989, quoted in *Inquiry*, Day 47, pp. 49–50.

19  DTI, *Memorandum*, 22 September 1989, quoted in *Inquiry*, Day 58, pp. 49–50.

20  *Inquiry*, Day 38, p. 143.

21  *Ibid*, Day 30, p. 63.

22  FCO, SEND, *Minute*, 28 September 1989, quoted in *Inquiry*, Day 22, pp. 93–5.

23  FCO, *Meeting with Lord Trefgarne and Mr Clark – Matrix Churchill and Guidelines on Arms Sales to Iran and Iraq*, 31 October 1989 (000301).

24  *Inquiry*, *op cit*, Evidence of Simon Sherrington, Day 40, pp. 56–62.

25  The Rt Hon. Sir Richard Scott, the Vice-Chancellor, *Report of the Inquiry into Exports of Defence Equipment and Dual-Use Goods to Iraq and Related Prosecutions*, House of Commons Papers 115, 1996, D6.151, p. 634.

26  *Ibid*, D8.13, p. 815.

27  *Ibid*, D5.35, p. 540.

28  SIS, *Telegrams*, 6 January 1989, (000113–14).

29  Paul Henderson, *Interview*, 22 September 1993.

30  *Inquiry*, *op cit*, Evidence of William Waldegrave, Day 30, p. 67.

31  *Ibid*, Evidence of Alan Clark, Day 50, p. 135.

32  SIS, *Minute* to the Cabinet Office, 24 July 1990, (000412–3).

33  *Minute* of the REU Meeting, 22 January 1988, quoted in *Inquiry*, *op cit*, Day 34, p. 103.

34  SIS, *Minute*, 8 July 1988, quoted in *Report*, *op cit*, D5.11, p. 515.

35  FCO, *Machine Tools for Iraq*, 28 January 1988, (000040–1).

36  DTI, *Matrix Churchill: Iraq*, Submission by the Head of OT2/3 to Private Secretary to Secretary of State, 26 September 1989, (000265).

37  Quoted in *Report*, *op cit*, D5.15, p. 519.

38  FCO, *Minute*, 29 June 1988, quoted in *Inquiry*, *op cit*, Day 19, pp. 84–5.

39  FCO, Minister of State, *Letter* to the Minister for Trade, 6 September 1989, (000223–4).

40  *Report*, *op cit*, D6.169, p. 643.

41  *Inquiry*, *op cit*, Evidence of Ian Mcdonald, Day 35, p. 32.

42  *Report*, *op cit*, D6.103, p. 609.

43  Quoted in *Report*, D6.101, p. 609 and D6.104, p. 610.

44  *Inquiry*, *op cit*, Day 43, p. 70; and Day 30 p. 99.

45  *Ibid*, Evidence of Rob Young, Day 23, pp. 56–8.

46  *Report*, *op cit*, D3.138–9, p. 433.

47  FCO, Minister of State, *Matrix Churchill: Export of Lathe Equipment to Iraq*, 6 February 1989, (000136).

48  FCO, SEND, *Minute*, 28 September 1989, quoted in *Inquiry*, *op cit*, Day 22, pp. 93–5.

49  DTI, *Summary Record of Ministerial Meeting to Decide Matrix Churchill Exports to Iraq*, 1 November 1989, (000308–9).

50  FCO, *Summary Record of the Ministerial Meeting to Decide Matrix Churchill Exports to Iraq*, 1 November 1989, (000306).

51  DTI, *Minute* of the Minister for Trade's Meeting with the MTTA, 20 January 1988, quoted in *Inquiry*, *op cit*, Day 46, p. 107.

52  Sir Derek Boorman, former Chief of Defence Intelligence, 1985–88, BBC, 11 February 1996.

53  *Report*, *op cit*, D8.6, p. 812.

54  FCO, *Summary Record of the IDC*, 22 January 1989, quoted in *Inquiry*, Day 47, pp. 152–3.

55  SIS, *Telegram*, 30 April 1990, (000364–5).

56  *Inquiry*, *op cit*, Day 48, p. 192.

57  JIC, *Report*, 12 July 1990, quoted in *Inquiry*, Day 35, p. 205.
58  *Ibid*, quoted in *Inquiry*, Day 35, p. 204.
59  *Report, op cit*, D8.11, p. 814
60  *Briefing for Ministerial Meeting*, 19 July 1990, the 'Iraq Note', 13 July 1990, quoted in *Inquiry, op cit*, Day 22, p. 21.
61  See for example: Minister for Defence Procurement, *Hansard*, 2 May 1990, col. 602; and anonymous officials quoted in *The Financial Times*, 13 September 1989.
62  *Minutes of the Ministerial Meeting*, 19 July 1990, quoted in *Inquiry, op cit*, Day 57, p. 48.
63  MOD, *Submission* to the MinDP, 21 September 1988, quoted in *Inquiry*, Day 29, p. 12.
64  *Inquiry*, Evidence of Alan Barrett, Day 38, p. 268.
65  MOD, *British Assistance to the Emerging Iraqi Arms Industry*, MODWG/DIS Survey, 29 October 1990, (000430).

**Chapter 9**

1  *Inquiry into Exports of Defence Equipment and Dual-Use Goods*, Hearings in the presence of the Rt Hon. Lord Justice Scott, Evidence of Ian McDonald, Day 35, p. 48.
2  *Ibid*, Evidence of Alan Clark, Day 50, p. 125.
3  *Ibid*, Evidence of Lt Col. Glazebrook, Day 4, p. 81.
4  Treasury Counsel, *Opinion*, 29 October 1990, quoted in the Rt Hon. Sir Richard Scott, the Vice-Chancellor, *Report of the Inquiry into the Export of Defence Equipment and Dual-Use Goods to Iraq*, House of Commons Papers 115, 1996, J1.21, p. 1607.
5  *Inquiry, op cit*, Evidence of Sir Hal Miller, Day 8, p. 47.
6  Anthony H. Cordesman and Abraham R. Wagner, *The Lessons of Modern War*, Vol. I, Boulder, Col., Westview Press, 1990, pp. 150–1.
7  MOD, DIS, *Loose Minute*, 23 December 1988, quoted in *Inquiry, op cit*, Day 22, p. 33.
8  MOD, DIS, *Loose Minute*, 23 December 1988, quoted in *Inquiry*, Day 22, p. 34; and MOD, *British Assistance to the Emerging Iraqi Arms Industry*, MODWG/DIS Survey, 29 October 1990, (000430)*.
    (* Numbered documents refer to those released to Paul Henderson and passed to the author.)
9  Trade and Industry Committee, *Second Report, Exports to Iraq: Project Babylon and Long-Range Guns*, Session 1991–92, House of Commons Papers 86, pp. xxi–xxiii.
10  See the MOD's assessment in *Ibid*, p. xxii.
11  Mohamed Heikal, *Illusions of Triumph: An Arab View of the Gulf War*, London, HarperCollins, 1992, p. 163.
12  Trade and Industry Committee, *Second Report, op cit*, HC 86, p. xv and Chris Cowley, Trade and Industry Committee, Minutes of Evidence, 15 January 1992, HC 86–iv, p. 218, QQ1117–1120.
13  FCO, *Memorandum*, 16 April 1987, quoted in *Inquiry, op cit*, Day 24, pp. 143–4.
14  FCO, *Memorandum*, 4 September 1986, quoted in *Inquiry*, Day 24, p. 127.
15  FCO, Minister of State, Comment, 8 August 1989, on a *Submission*, 4 August 1989, quoted in *Inquiry*, Day 26, p. 15.
16  *Inquiry*, Day 16, p. 96.
17  See for example: *Inquiry*, Evidence of Eric Beston, Day 43, pp. 131–2.
18  William Lowther, *Arms and the Man: Dr Gerald Bull, Iraq and the Supergun*, London, Macmillan, 1991, pp. 238–9.
19  *Inquiry, op cit*, Evidence of Timothy Renton MP, Day 24, p. 141.
20  DTI, *Matrix Churchill: Iraq, Meeting with Mr Waldegrave and Mr Clark, 1 November*, 31 October 1989, (000290).
21  See, for example, *Inquiry, op cit*, Evidence of Eric Beston, Day 44, p. 12.
22  *Ibid*, Evidence of Lord Trefgarne, Day 82, p. 13.
23  *Ibid*, Evidence of Sir David Miers, Day 19, p. 52.
24  *Hansard*, 11 January 1993, col. 555; and 19 January 1993, col. 650.
25  *Memorandum* submitted by the DTI, (EQ6), Trade and Industry Committee, *Exports to Iraq*, Memoranda of Evidence, 17 July 1991, Session 1990–91, House of Commons Papers 607, p. 23.

26 *Briefing for Ministerial Meeting*, 19 July 1990, the 'Iraq Note', 13 July 1990, quoted in *Report, op cit*, D3.155, p. 441.

27 FCO, *Submission* to Secretary of State, 6 October 1989, quoted in *Inquiry, op cit*, Day 22, p. 115.

28 *Ibid.*

29 Alan Clark, *Diaries*, London, Wiedenfeld and Nicolson, 1993, p. 296.

30 Trade and Industry Committee, *Second Report, op cit*, HC 86, p. xv.

31 *Memorandum* submitted by the DTI, (EQ6), *op cit*, Trade and Industry Committee, HC 607, p. 23.

32 *The Guardian*, 23 November 1992.

33 Department of Energy, *UK Energy Sector Well Placed for Trade with Iraq, Press Release 156*, 13 October 1989, (000277).

34 CIA, *World Factbook*, quoted in Anthony H. Cordesman, *After the Storm: The Changing Military Balance in the Middle East*, London, Mansell Publishing Ltd, 1993, p. 436.

35 CIA, Directorate of Intelligence, *Iraq: No End in Sight to Debt Burden*, 12 April 1990.

36 *Inquiry, op cit*, Evidence of Sir David Miers, Day 18, p. 135.

37 CIA, *Iraq: No End in Sight, op cit.*

38 *Ibid.*

39 Treasury, *Letter*, 20 March 1990, Foreign Secretary, *Letter*, 27 March 1990, Secretary of State for Defence, *Letter*, 30 March 1990, quoted in *Report, op cit*, D3.182, pp. 464–5.

40 Secretary of State for Trade and Industry, *Memorandum*, 21 June 1990, quoted in *Inquiry, op cit*, Day 58, p. 77.

41 Secretary of State for Trade and Industry, *Trade with Iraq*, Letter to the Prime Minister, 21 June 1990, (000400 and 000402).

42 *Inquiry, op cit*, Evidence of William Waldegrave, Day 30, pp. 16–18; and FCO, Minister of State, *Matrix Churchill: Export of Lathe Equipment to Iraq*, 6 February 1989, (000136).

43 FCO, Minister of State, Note on *Submission*, 6 October 1989, quoted in *Inquiry*, Day 17, p. 11.

44 DTI, *Matrix Churchill Ltd: Export Licence Applications for Iraq*, 25 September 1989, (000264).

45 Secretary of State for Trade and Industry, *Trade with Iraq*, Letter to the Prime Minister, 21 June 1990, (000401); and *Briefing for Ministerial Meeting*, 19 July 1990, the 'Iraq Note', 13 July 1990, quoted in *Inquiry, op cit*, Day 22, p. 21.

46 *Briefing for Ministerial Meeting*, 19 July 1990, the 'Iraq Note', quoted in *Inquiry*, Day 22, p. 21.

47 DTI, *Memorandum* to the Secretary of State for Trade and Industry, 14 June 1990, (000396).

48 Yezid Sayigh, *Arab Military Industry: Capability, Performance and Impact*, London, Brassey's, 1992, p. 23.

49 Subcommittee on International Economic Policy and Trade of the Committee on Foreign Affairs, House of Representatives, *United States Export of Sensitive Technology to Iraq*, Hearings, Hon. Sam Gejdenson, Chairman, 8 April 1991, 102nd Congress, First Session, Washington DC, USGPO, 1991, pp. 1–2.

50 *Statement* of Henry B. Gonzalez, Chairman Committee on Banking, Finance and Urban Affairs, 21 January 1993.

51 *Inquiry, op cit*, Evidence of Mr Petter, Day 58, p. 96.

52 DTI, *Matrix Churchill Ltd, Iraq: Proposed Meeting with MD*, 20 September 1989, (000241).

53 Foreign Secretary, *Letter* to the Home Secretary, Summer 1988, quoted in *Inquiry, op cit*, Day 19, p. 103.

54 FCO, SEND, *Memorandum*, 28 September 1989, quoted in *Inquiry*, Day 22, pp. 93–4.

55 DTI, *Speaking Note for Meeting of MISC 118*, 27 June 1990, quoted in *Inquiry*, Day 58, p. 90.

56 J. B. Kelly, *Arabia, the Gulf and the West*, London, HarperCollins, 1980, p. 308.

57 Anthony H. Cordesman, 'Arms to Iran: The Impact of US and Other Sales on the Iran–Iraq War', *American-Arab Affairs*, Vol. 20, Spring 1987, p. 25.

58  SIS, *Telegram*, 18 January 1988, (000029).
59  MTTA, *Report of Meeting with Alan Clark MP, Minister for Trade, Concerning Export Licences for Iraq*, 20 January 1988.
60  BBC, 'News West', 24 June 1992.
61  *Hansard*, 24 November 1992, col. 609.
62  DTI, *Submission* to Minister for Trade, 1 June 1990, quoted in *Inquiry, op cit*, Day 82, pp. 145–7.
63  Greg Thompson, quoted in *The Guardian*, 19 July 1995.
64  *The Observer*, 4 February 1990.
65  FCO, Minister of State, *Letter* to Minister for Trade, 27 April 1898, (000170).
66  *Hansard*, 1 March 1993, col. WA26.

## Conclusion

1  Bernard Porter, *Britain, Europe and the World, 1850–1986: Delusions of Grandeur*, Second Edition, London, George Allen and Unwin, 1989, p. 146.
2  L. Festinger, *A Theory of Cognitive Dissonance*, Evanston, Ill., Row Peterson, 1957.
3  Edward A. Kolodziej, *Making and Marketing Arms: the French Experience and its Implications for the International System*, Princeton, NJ, Princeton University Press, 1987.
4  *Inquiry into Exports of Defence Equipment and Dual Use Goods to Iraq*, Hearings in the presence of the Rt Hon. Lord Justice Scott, Evidence of David Mellor, MP, Day 25, p. 58.
5  Christian Catrina, *Arms Transfers and Dependence*, New York, UN Institute for Disarmament Research, Taylor & Francis, 1988, p. 326.

# Select bibliography

### Documentary sources

**United Kingdom**

*Inquiry into Exports of Defence Equipment and Dual-Use Goods to Iraq*, Hearings in the presence of the Rt Hon. Lord Justice Scott, Transcripts of the computerized stenograph notes of Smith Bernal Reporting Ltd.

The Rt Hon. Sir Richard Scott, the Vice-Chancellor, *Report of the Inquiry into the Export of Defence and Dual-Use Goods to Iraq and Related Prosecutions*, House of Commons Papers 115, 1996.

Secretary of State for Foreign and Commonwealth Affairs, *Foreign Affairs Committee Report, Current United Kingdom Policy Towards the Iran/Iraq Conflict*, Cm 505, October 1988.

Department of Trade and Industry, *Exports to Iraq: Project Babylon and Large Range Guns*, Government's Response to the Second Report of the Trade and Industry Committee in Session 1991–92, Cm 2019.

Foreign Affairs Committee, *Second Report, Current UK Policy towards the Iran/Iraq Conflict*, Report together with Minutes of Evidence with Appendices, Session 1987–88, House of Commons Papers 279–I and 279–II.

Foreign Affairs Committee, *Overseas Arms Sales: Foreign Policy Aspects*, Minutes of Evidence, 4 March 1981, Session 1980–81, House of Commons Papers 41–iv.

National Audit Office, Report by the Comptroller and Auditor General, *Ministry of Defence: Support for Defence Exports*, Session 1988–89, House of Commons Papers 303.

Committee of Public Accounts, *Fortieth Report, Ministry of Defence: Support for Defence Exports*, together with Minutes of Evidence, Session 1988–89, House of Commons Papers 339–i.

Trade and Industry Committee, *Exports to Iraq*: Memoranda of Evidence, 17 July 1991, Session 1990–91, House of Commons Papers 607.

Trade and Industry Committee, *Second Report, Exports to Iraq: Project Babylon and Long-Range Guns*, together with Minutes of Evidence, Session 1991–92, House of Commons Papers 86–xv.

Trade and Industry Committee, *Export Licensing and BMARC*, Minutes of Evidence, Session 1995–96, House of Commons Papers 87–i to iii.

**United States of America**

Subcommittee on International Economic Policy and Trade of the Committee on Foreign Affairs, House of Representatives, *United States Exports of Sensitive Technology to Iraq*, Hearings, 8 April 1991, 102nd Congress, First Session, Washington DC, US PO, 1991.

United States House of Representatives Select Committee to Investigate Covert Arms Transactions with Iran; US Senate Select Committee on Secret Military Assistance to Iran and the Nicaraguan Opposition, *Report of the Congressional Committees Investigating the Iran-Contra Affair*, H. Rept. No. 100–433, 100th Congress, First Session, S. Rept. No. 100–216, Washington DC, US GPO, 1987.

## Books

Anthony, I. (Ed.), *Arms Export Regulations*, New York, Oxford University Press, SIPRI, 1991.

Axelgard, Frederick W. (Ed.), *Iraq in Transition: A Political, Economic and Strategic Perspective*, Boulder, Colorado, Westview Press, 1986.

Axelrod, R., *The Structure of Decision: The Cognitive Maps of Elites*, Princeton, Princeton University Press, 1976.

Bakhash, S., *The Reign of the Ayatollahs: Iran and the Islamic Revolution*, London, Castlepoint, 1986.

Bartlett, C. J., *British Foreign Policy in the Twentieth Century*, Basingstoke, Macmillan, 1989.

Brenchley, F., *Britain and the Middle East: An Economic History, 1945–1987*, London, Crook Academic Publishers, 1989.

Brzoska, Michael, *International Arms Transfers: The Nature and Dimension of the Problem*, Hamburg, University of Hamburg, April 1990.

Brzoska, Michael and Ohlson, Thomas, *Arms Production in the Third World*, Taylor and Francis, London and Philadelphia, 1986.

Byrd, P. (Ed.), *British Foreign Policy Under Thatcher*, Oxford, Philip Allan, 1988.

Cannizzo, C. (Ed.), *The Gun Merchants: Politics and Policies of the Major Arms Suppliers*, Oxford, Pergamon Press, 1980.

Catrina, Christian, *Arms Transfers and Dependence*, UN Institute for Disarmament Research, New York, Taylor & Francis, 1988.

Chubin, Shahram & Tripp, Charles, *Iran and Iraq at War*, London, T.B. Tauris, 1988.

Clark, Alan, *Diaries*, London, Weidenfeld and Nicolson, 1993.

Clarke, Michael, *British External Policy-Making in the 1990s*, London, Macmillan for the Royal Institute of International Affairs, 1992.

Clarke, Michael and White, Brian (Eds), *Understanding Foreign Affairs: The Foreign Policy Systems Approach*, Aldershot, Edward Elgar, 1989.

Committee Against Repression and For Democratic Rights in Iraq, *Saddam's Iraq: Revolution or Reaction?*, 2nd Edition, London, Zed Books Ltd, 1989.

Cordesman, Anthony H., *After the Storm: The Changing Military Balance in the Middle East*, London, Mansell Publishing Ltd, 1993.

Cordesman, Anthony H., *The Gulf and the West: Strategic Relations and Military Realities*, Boulder, Colorado, Westview Press, 1988.

Cordesman, Anthony, H. and Wagner, Abraham R., *The Lessons of Modern Wars: The Iran–Iraq War, Vol.II*, Boulder, Colorado, Westview Press; London, Mansell Publishing, 1990.

Cottam, R., *Foreign Policy Motivation: A General Theory and a Case Study*, Pittsburgh, University of Pittsburgh Press, 1977.

Darby, P., *British Defence Policy East of Suez, 1947–68*, London, Oxford University Press, 1973.

Frankel, J., *British Foreign Policy, 1945–73*, London, Oxford University Press, 1975.

Freedman, Lawrence and Clarke, Michael (Eds), *Britain in the World*, Cambridge, Cambridge University Press, 1991.

Freedman, Lawrence, *Arms Production in the United Kingdom: Problems and Prospects*, London, Royal Institute of International Affairs, 1978.

Gamble, Andrew, *Britain in Decline: Economic Policy, Political Strategy and the British State*, Fourth Edition, London, Macmillan, 1994.

Hammond, P., Louscher, D., Salomone, M. and Grahm, N., *The Reluctant Supplier: US Decision-Making for Arms Sales*, Cambridge, Mass., Oelgeschlager, Gunn and Hain, 1983.

Harkavy, Robert E., *The Arms Trade and International Systems*, Cambridge, Mass., Ballinger Publishing Company, 1975.

Harkavy, Robert E. and Neuman, Stephanie G. (Eds), *The Annals of the American Academy of Political and Social Science*, The Arms Trade: Problems and Prospects in the Post-Cold War World, London, Sage, 1994.

Henderson, Paul, *The Unlikely Spy*, London, Bloomsbury, 1993.

Howe, Geoffrey, *Conflict of Loyalty*, London, Macmillan, 1994.

Hutton, Will, *The State We're In*, London, Vintage, 1996.

James, Gerald, *In the Public Interest*, London, Little, Brown and Co., 1995.

Janis, I., *Victims of Groupthink*, Boston, Houghton-Mifflin, 1972.

Jervis, R., *Perception and Misperception in International Politics*, Princeton, Princeton University Press, 1976.

Karsh, Efraim, Navias, M. and Sabin, P. (Eds), *Non-Conventional Weapons Proliferation in the Middle East*, Oxford, Oxford University Press, 1992.

Karsh, Efraim (Ed.), *The Iran–Iraq War: Impact and Implications*, London, Macmillan Press, 1989.

Kelly, J. B., *Arabia, the Gulf and the West*, London, Basic Books, HarperCollins, 1980.

Kent, M., *Moguls and Mandarins: Oil Imperialism and the Middle East in British Foreign Policy*, London, Frank Cass, 1993.

Khadduri, Majid, *The Gulf War: The Origins and Implications of the Iraq–Iran Conflict*, Oxford, Oxford University Press, 1988.

Kolodziej, Edward A., *Making and Marketing Aims: The French Experience and its Implications for the International System*, Princeton, Princeton University Press, 1987.

Kornbluh, Peter and Byrne, Malcolm (Eds), *The Iran-Contra Scandal: The Declassified History*, New York, The New Press, 1993.

Krause, Keith, *Arms and the State: Patterns of Military Production and Trade*, Cambridge, Cambridge University Press, 1992.

Laurance, Edward J., *The International Arms Trade*, New York, Lexington, Macmillan, 1992.

Lowther, William, *Arms and the Man: Dr Gerald Bull, Iraq and the Supergun*, London, Macmillan, 1991.

The National Security Archive, *The Chronology: The Documented Day to Day Account of the Secret Military Assistance to Iran and the Contras*, New York, Warner Books, 1987.

Neuman, Stephanie G. and Harkavy, Robert E. (Eds), *Arms Transfers in the Modern World*, New York, Praeger, 1979.

Northedge, F. S., *Descent from Power: British Foreign Policy, 1945–1973*, London, Allen and Unwin, 1974.

O'Ballance, Edgar, *The Gulf War*, London, Brassey's, 1988.

Ohlson, Thomas (Ed.) (SIPRI), *Arms Transfer Limitations and Third World Security*, Oxford/New York, Oxford University Press, 1988.

Pierre, Andrew J., *The Global Politics of Arms Sales*, Princeton NJ, Princeton University Press, 1982.

Porter, Bernard, *Britain, Europe and the World, 1850–1986: Delusions of Grandeur*, Second Edition, London, George Allen and Unwin, 1989.

Pridham, Brian R. (Ed.), *The Arab Gulf and the West*, London, Croom Helm, 1985.

Sampson, Anthony, *The Arms Bazaar*, Second Edition, London, Hodder and Stoughton, 1988.

Sanders, David, *Losing an Empire, Finding a Role: British Foreign Policy Since 1945*, London, Macmillan, 1990.

Sayigh, Yezid, *Arab Military Industry: Capability, Performance and Impact*, London, Brassey's, 1992.

Shultz, G. P., *Turmoil and Triumph: My Years as Secretary of State*, New York, Scribners, 1993.

Smith, Michael, Smith, Steve and White, Brian (Eds), *British Foreign Policy: Tradition, Change and Transformation*, London, Unwin Hyman, 1988.

Smith, Steve and Clarke, Michael (Eds), *Foreign Policy Implementation*, London, George Allen and Unwin, 1985.

Spiers, Edward M., *Chemical and Biological Weapons: A Study of Proliferation*, London, Macmillan, 1994.

Stanley, J. and Pearton, M., *The International Trade in Arms*, London, Chatto and Windus, 1972.

Stockholm International Peace Research Institute, *World Armaments and Disarmament:*

*SIPRI Yearbook,* London and Philadelphia, Taylor and Francis, and Oxford, Oxford University Press, various years.

Stockholm International Peace Research Institute, *The Arms Trade with the Third World,* London, Paul Elek, 1971 and Harmondsworth, Penguin Books Ltd, 1975.

Timmerman, Kenneth R., *The Death Lobby: How the West Armed Iraq,* London, Fourth Estate, 1992.

*The Tower Commission Report,* New York, Bantam Books and Times Books, 1987.

Tugendhat, C. and Wallace, W., *Options for British Foreign Policy in the 1990s,* London, Routledge/Royal Institute of International Affairs, 1988.

Wallace, W., *The Foreign Policy Process in Britain,* London, Royal Institute of International Affairs, 1975.


## Articles

Axelgard, F. W., 'Iraq: The Post-War Setting', *American-Arab Affairs* Vol. 28, Spring 1989.

Baranson, Jack, 'Technology Exports Can Hurt Us', *Foreign Policy* No. 25, Winter 1976–77.

Blackaby, Frank and Ohlson, Thomas, 'Military Expenditure and the Arms Trade: Problems of Data', *Bulletin of Peace Proposals* Vol. 13, No. 4, 1982.

Boulding, Kenneth E., 'National Images and International Systems', *Journal of Conflict Resolution* Vol. 30, 1959.

Brecher, Michael, Steinberg, B. and Stein, J., 'A Framework for Research on Foreign Policy Behaviour', *Journal of Conflict Resolution* Vol. 13, 1969.

Cahn, Anne Hessing, 'Arms Sales Economics: The Sellers' Perspectives', *Stanford Journal of International Studies* Vol. 14, Spring 1979.

Cordesman, Anthony H., 'Arms to Iran: The Impact of US and Other States on the Iran–Iraq War', *American-Arab Affairs* Vol. 20, Spring 1987.

Cornford, J. P., 'Review Article: The Illusion of Decision', *British Journal of Political Science* Vol. 4, 1974.

Edmonds, Martin, 'The British Government and Arms Sales', *Arms and Disarmament Information Unit Report*, Vol. IV, No. 6, November/December 1982.

Freedman, Lawrence, 'The Arms Trade: A Review', *International Affairs,* London, Vol. 55 No. 3, July 1979.

Freedman, Lawrence, 'British Foreign Policy to 1985: IV Britain and the Arms Trade', *International Affairs* Vol. 54 No. 3, July 1978.

Freedman, Lawrence, 'Logic, Politics and Foreign Policy Processes: A Critique of the Bureaucratic Politics Model', *International Affairs* Vol. 52 No. 3, 1976.

Gamble, Andrew, 'Theories of British Politics', *Political Studies* Vol. 38, 1990.

Gelb, Leslie, 'Arms Sales', *Foreign Policy* Vol. 25, 1976.

Griffin, J. C. and Rouse, W., 'Countertrade as a Third World Strategy of Development', *Third World Quarterly* Vol. 8 No. 1, January 1986.

Harkavy, Robert E., 'Arms Resupply During Conflict: A Framework for Analysis', *The Jerusalem Journal of International Relations* Vol. 7 No. 3, 1985.

Howe, G., 'Sovereignty and Interdependence: Britain's Place in the World', *International Affairs* Vol. 66 No. 4, 1990.

Hutton, W., 'Britain in a Cold Climate: The Economic Aims of Foreign Policy in the 1990s', *International Affairs* Vol. 68 No. 4, 1992.

Julius, D., 'Britain's Changing International Interests: Economic Influences on Foreign Policy Priorities', *International Affairs* Vol. 63 No. 3, 1987.

King, Ralph, 'The Iran–Iraq War: The Political Implications', *Adelphi Papers* Vol. 219 No. 115, London, 1987.

Klare, Michael T., 'Secret Operatives, Clandestine Trades: The Thriving Black Market for Weapons', *Bulletin of the Atomic Scientists*, April 1988.

Klare, Michael T., 'The Arms Trade: Changing Patterns in the 1980s', *Third World Quarterly* Vol. 9 No. 4, October 1987.

Klare, Michael T., 'The Unnoticed Arms Trade: Exports of Conventional Arms-Making Technology', *International Security* Vol. 8 No. 2, Fall 1983.

Laurance, Edward J., 'The New Gunrunning', *Orbis* Vol. 33 No. 2, Spring 1989.

McNaugher, Thomas L., 'Ballistic Missiles and Chemical Weapons: The Legacy of the . Iran–Iraq War', *International Security* Vol. 15 No. 2, Fall 1990.

Moberly, J., 'Iraq in the Aftermath of the Gulf War', *Asian Affairs* Vol. 76 No. 3, October 1989.

Navias, M., 'Ballistic Missile Proliferation in the Third World', *Adelphi Paper* No. 252, London, Brassey's for IISS, 1990.

Neuman, Stephanie G., 'Co-production, Barter and Countertrade Offsets in the International Arms Market', *Orbis* Vol. 29 No. 1, 1985.

Nolan, Janne E., 'Ballistic Missiles in the Third World – The Limits of Non-Proliferation', *Arms Control Today*, November 1989.

Pearson, Frederic S., 'The Question of Control in British Defence Sales Policy', *International Affairs* Vol. 59 No. 2, 1983.

Pierre, Andrew J., 'Arms Sales: The New Diplomacy', *Foreign Affairs* Vol. 60 No. 2, 1981.

Ross, Andrew L., 'Do-It-Yourself Weaponry', *Bulletin of Atomic Scientists,* May 1990.

Smith, Ron, Humm, Anthony and Fontanel, Jacques, 'The Economics of Exporting Arms', *Journal of Peace Research* Vol. 22 No. 3, 1985.

Taylor, Trevor, 'Research Note: British Arms Exports and Research and Development Costs', *Survival* Vol. XXII, No. 6, November/December 1980.

Tripp, Charles, 'Europe and the Persian Gulf', *Cambridge Review of International Affairs* Vol. 2 No. 2, 1988.

Vayrynen, Raimo, 'Economic and Political Consequences of Arms Transfers to the Third World', *Alternatives* Vol. 6 No. 1, March 1980.

## Papers and pamphlets

Campaign Against the Arms Trade, *Steps to Controlling and Ending the Arms Trade: A Policy Outline for a Future British Government*, CAAT, London, 1991.

Campaign Against the Arms Trade, *The Government and the Arms Trade*, London, CAAT, 1990.

McGowan, J., 'The Future for UK Defence Exports', *Seaford House Papers*, Royal College of Defence Studies, London, 1986.

Oxford Research Group, *International Control of the Arms Trade,* Oxford, Oxford Research Group, April 1992.

Saferworld Foundation, *Regulating Arms Exports: a Programme for the European Community,* Bristol, Saferworld Foundation, September 1991.

Simpson, John (Ed.), *The Control of Arms Transfers: Report of a FCO/BISA Seminar*, 23 September 1977 London, Arms Control and Disarmament Research Unit, Foreign and Commonwealth Office.

Svenska Freds-Och Skiljedomsforeningen (The Swedish Peace and Arbitration Society, SPAS), *The International Connections of the Bofors Affair,* Stockholm, SPAS, December 1987.

World Development Movement, *Gunrunners' Gold*, London, World Development Movement, 1995.

# Appendix A:

## The explosives cartel – shipping documents

000072

## Zeebrugge-Mole
### Manifest of goods

| Shippers | Consignees | Marks and numbers | Number | Package Type | Merchandise | Gross weight kilos | Mesure | Remarks |
|---|---|---|---|---|---|---|---|---|
| ROYAL ORDNANCE EXPLOSIVES BRIDGWATER UNITED KINGDOM | ELTREME S.A. ATHENA CENTER 32 KIFISSIAS AVE PARADISSOS AMAROUSSION ATHENS - GREECE | ORDER N°10941/1-1401-16531-11 UR SERIAL N°0208  1.9.0 BOMB BLAST SIGN | 104 1 | CASES CASE | 2.362 KG NETT TETRYL LOADED INTO 1X20'CTR N°2101 "SAMPLE | 2.492 | 2.120 M3 | 1X0 1 DIV. 1.1.C UNN°0208 FREIGHT PREPAID |
| NITROX CHEMIE HOLLAND | IDEM | CTR NOS. 100441-7/216 DRUMS 244155-7/282 DRUMS 108814-0/772 DRUMS | 500 | DRUMS | 40.000 KG NETT SMOKELESS POWDER LOADED INTO 1X20'CTRS | 49.645 TARRA CTRS INCLUDED | | 1X0 1 DIV. 1.3.C UNN°0161 FREIGHT PREPAID |
| P.R.B. GECREN BRUSSELS BELGIUM | IDEM | PRIMER BM 139 B1 L/C N°1618627/6  CODE N°01 REF.N°30847/1-1401-16102/59 BALL POWDER FOR CARTRIDGES CALIBRE 7.62MM/5,5IP CL 1/46 L/C N°1618627/6 REF. 30847/1-1401-16011.55 | 60 2.000 | CASES DRUMS | 1°.44 PRIMER BM139B1 LOT PRB-01-01 LOADED INTO 1X20'CTR N°2109 100.000 KG BALL POWDER FOR CARTRIDGES CALIBRE 7.62MM LOADED INTO 1X20'CTRS N°2109 | 2.714 11.000 | 4.011 M3 167.416 M3 | 1X0 1 DIV. 1.1.D UNN°0319 FREIGHT COLLECT 1X0 1 DIV. 1.3.C UNN°0161 FREIGHT COLLECT |
| IDEM | IDEM | PRIMER BM 114 B1 L/C N°1618627/6  CODE N°01 REF.N°30847/1-1401-16102/55 | 45 | CASES | 11.505 PRIMERS BM114B1 LOT PRB-01-01 LOADED INTO 1X20'CTR N°2109 | 2.714 | | 1X0 1 DIV. 1.3.C UNN°0319 FREIGHT COLLECT |

Ship's ......3-M...
...... Manifest of goods shipped on board at the ...DANISA...

...... Captain ...... J. NIELSEN .......  ... at the ...DANISA...

DATE JUNE 1986 ...... JOTEH

Destination RAPHAN ALPAS

| Shippers | Consignees | Marks and numbers | Packages and merchandise | | | Gross weight | Said | Remarks |
|---|---|---|---|---|---|---|---|---|
| | | | Number | Package type | Merchandise | Price / Kilo | | |
| 18/1 | ROYAL ORDNANCE EXPLOSIVES ASSOCIATES UNITED KINGDOM | TO ORDER OF BANK NAT-EAST-IRAN-TEHRAN-IRAN NOTIFY : SASEMAHE SANA-TE DEFA'E JOMHOURI ESLAMI IRAN | ORDER N° 30847/1-1601-30535-18 UN SERIAL N°0209 1.1.D BOMB BLAST SIGN | 104 1 | CASES CASE "SAMPLE" | 7,161 KG NETT TOTAL LOADED INTO 1X20'CTR N°2101 | 7,493 | 12no | INS I DEV 5.A.P HNN°CGN1 FREIGHT PREPAID |
| 18/2 | RHEDEE CHEMIE HOLLAND | IDEM | CTR.NOS 10641/1-7/266 DRUMS 24453H-3/282 DRUMS 10641A-0/237 DRUMS | 800 | DRUMS | 40.0MM AL NETT SMOKELESS POWDR LOADED INTO 1X20'CTRS | 49.643 TARA CTRS INCLUDED | | INS I DEV 1 I HNN°CGN1 340 NETT PREPAID |
| 18/3 | F.N. GEORGE BRUSSELS BELGIUM | IDEM | PRIMER DM 131 B) L/C N°161847/6 CODE N°01 REF.N°10847/1-1601-16707/17 BALL POWDER FOR CARTRIDGES CALIBRE 7.62NP/LGY CT.1/144 L/C N°30189A/R REF N°10842/1-1601-1072° 21 | 40 2,393 | CASES DRUMS | 11.TWO PRIMERS DM131B) LOT PRD-00-07 LOADED IX-0 N°2201/13 N°2101 1940.UNO AL POIL PUMOID SAN. CARTRIDGES CALIBER 1.26N LOADED INTO N°101'CTR. 2100-31WEON 4.104/PUWOR 2100/231PM 2100/13WOR | 7.76 131 | 4 issued | INS I DEV 1 I HNN°GCN-1 1no 27 5.4 6.0 |
| 18/4 | IDEM | IDEM | PRIMER DM 131 B) L/C N°161847/6 CODE N°03 REF.N°30842/1-1601-16707/14 | 40 | CASES | 13.BOM PRIMER DM131B) LOT PRD-00-07 LOADED INTO N°2101. | 224 | | INS I DEV 1 I HNN°CGN1 PREIGHT INCL |

# Appendix B:

Secret Intelligence Service, Iraq: Project 'Babylon' to develop the technology for Hypervelocity Gun with extreme range capability (October–November 1989)

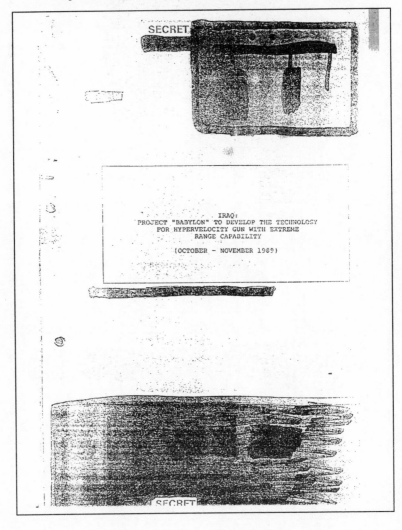

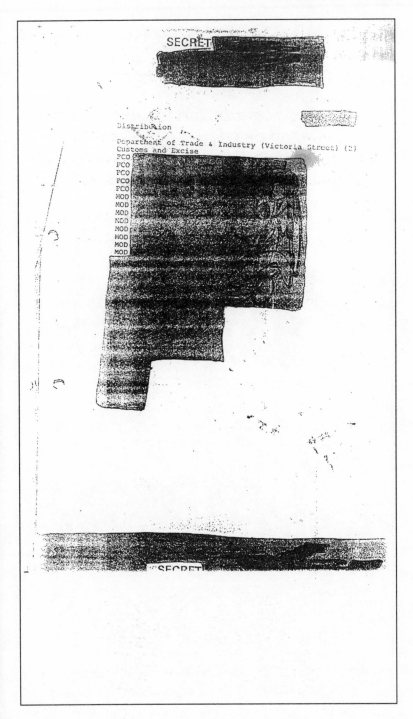

SECRET

Distribution

Department of Trade & Industry (Victoria Street) (2)
Customs and Excise
FCO
FCO
FCO
FCO
FCO
MOD
MOD
MOD
MOD
MOD
MOD
MOD
MOD

SECRET

SECRET

Report

Title: IRAQ: PROJECT "BABYLON" TO DEVELOP THE TECHNOLOGY FOR HYPERVELOCITY GUN WITH EXTREME RANGE CAPABILITY

Date of
Information: October-November 1989

Source: Multiple

1. The Iraqi Ministry of Industry is, under Project BABYLON, endeavouring to acquire the technology for a hypervelocity gun capable of delivering substantial payloads to extreme ranges. There is no indication of the intended operational role of such a weapon but it would seem most suited for long range strategic bombardment of wide area targets such as cities.

2.

3. At a later stage in the project, a larger operational version of the gun will fire 262 mm projectiles with flip out fins. This will be a Rocket Assisted Projectile (RAP) and carry a warhead section of 1.28 m length. This is unlikely to be a simple High Explosive (HE) device and is most likely a carrier arrangement for Cluster Munitions though Fuel Air Explosive (FAE) or Chemical Warfare (CW) payloads cannot be ruled out. The project specification calls for it to be able to withstand a considerable amount of kinetic heating (3000°C for 10 minutes). While it would be unwise to use this to read back to a velocity and ultimately a range requirement (safety factors being unknown), it is clear that ranges of several hundred kilometers are contemplated. A figure of 600 km has been mentioned.

4.

SECRET

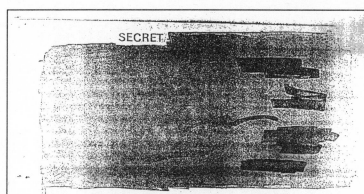

SECRET

5.  Drawings are attached showing:

   a

   b

   c  elements of the larger operational projectile.

Desk Comment

1.  We do not underestimate the difficulties of developing such a weapon to the point where it gives a meaningful operational capability. However the Iraqis would appear to have adopted a proper systematic approach ▓▓▓▓▓▓▓ ▓▓▓▓▓▓▓▓▓▓▓ which could lead to a successful outcome given continued access to the necessary resources and West European technology.

2.  The weapon is not dissimilar in concept to the V3 developed by the Germans towards the end of WWII. Similar concepts have been described by Dr BULL in some of his published works for a range of application including that of a Space Launch Vehicle.

3.  The technology required to achieve this project would not appear to be constrained by the terms of the Missile Technology Control Regime (MTCR) which may be part of the motivation for the project given the difficulties of acquiring ballistic missile technology.

SECRET

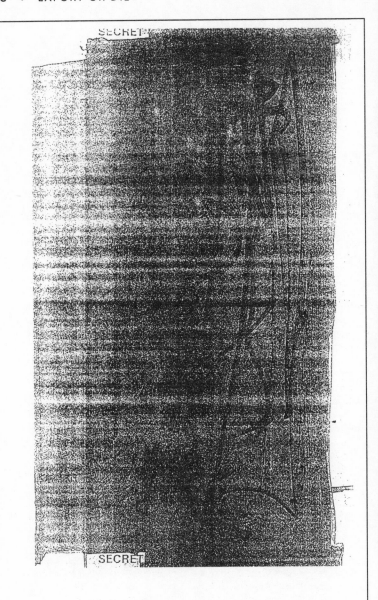

# Index